W9-BZL-937

BEYOND THE GOD PARTICLE

BEYOND THE GOD PARTICLE

LEON LEDERMAN

NOBEL LAUREATE

CHRISTOPHER HILL

Prometheus Books

59 John Glenn Drive
Amherst, New York 14228–2119

Published 2013 by Prometheus Books

Cover design by Grace M. Conti-Zilsberger

Inquiries should be addressed to
Prometheus Books
59 John Glenn Drive
Amherst, New York 14228–2119
VOICE: 716–691–0133
FAX: 716–691–0137
WWW.PROMETHEUSBOOKS.COM

17 16 15 14 13 5 4 3 2 1

Library of Congress Cataloging-in-Publication Data

Lederman, Leon M., author.
Beyond the god particle / Leon M. Lederman, Christopher T. Hill.
pages cm
Includes bibliographical references and index.
ISBN 978-1-61614-801-0 (hardback)
ISBN 978-1-61614-802-7 (ebook)
1. Higgs bosons. 2. Particles (Nuclear physics)—Philosophy. 3. Matter—Constitution. I. Hill, Christopher T., 1951- author. II. Title.

QC793.5.B62L425 2013
539.7′2—dc23

2013022346

Printed in the United States of A

We dedicate this book to our fellow citizens
who through their taxes have graciously supported basic research.

CONTENTS

ACKNOWLEDGMENTS

We thank our dear late editor, Linda Greenspan Regan, for her tireless efforts in initiating this project, and her excellent editorial skills throughout our previous projects, *Symmetry and the Beautiful Universe* and *Quantum Physics for Poets.* We also thank Julia DeGraf, Jill Maxick, Brian McMahon, Steven L. Mitchell, and Grace M. Conti-Zilsberger, who have served as our editors through the oft-time hectic completion of this book. For useful comments and advice, we thank Ronald Ford, William McDaniel, Ellen Lederman, and especially Maureen McMurrough and her dachshunds.

The authors also underscore the importance of the learning of science by young people. Here, the continuing efforts of Prometheus Books in the publication of science books are gratefully acknowledged, as are the huge contributions of schools throughout the country. We especially acknowledge the Illinois Mathematics and Science Academy, recognized as one of the most successful science high schools in the nation, and Fermilab, the greatest national laboratory in the Western Hemisphere and our favorite in the solar system.

Out of the mid-wood's twilight
Into the meadow's dawn,
Ivory limbed and brown-eyed,
Flashes my Faun!
He skips through the copses singing,
And his shadow dances along,
And I know not which I should follow,
Shadow or song!
O Hunter, snare me his shadow!
O Nightingale, catch me his strain!
Else moonstruck with music and madness
I track him in vain!

—Oscar Wilde, "In the Forest"

CHAPTER 1

INTRODUCTION

There is a serene valley stretching about Lac Léman, from the spectacular French Alps, with Mont Blanc looming in the distance to the east, and the ancient round-top Jura Mountains to the west. This is the countryside of France surrounding Geneva, Switzerland, one of the most beautiful regions in all of Europe. It is a historic place; a center of banking and watchmaking; the home of Voltaire, the grandmaster philosopher of the Enlightenment; the host to the League of Nations, forerunner of the United Nations; a mecca for Roman history buffs, foodies, skiers, and train watchers. Geneva is a French-speaking canton of Switzerland, on the border with France, not more than a few kilometers from downtown.

Within the environs of Geneva we find, today, thousands of the world's best-trained and most highly educated physicists hard at work. In a metaphorical sense, they have become the enchanted dwarves, the "Nibelungen" of the ancient Norse pagan religion who lived in a mystical region called Nibelheim, where they mined and toiled in the bowels of the earth. They ultimately created, from the gold of the Rhine, a ring of enormous magical power for whoever possessed it. It is here in Geneva that the metaphorical Nibelungen, the modern-day particle physicists, have fabricated, deep in the bowels of the earth, a mighty ring of their own.

The physicists are real and their ring is real. It is not made of gold but rather of tons of steel, copper, aluminum, nickel, and titanium, enclosed within enormous vats of super-cooled liquid helium, decorated with the most resplendent and powerful electronics to be found on Planet Earth. With the mighty ring near Geneva, some 300 feet underground, the scientists painstakingly toil hard into the early morning hours. The product of their labor is not gold but the creation of a new form of matter, far, far more valuable than gold and never before seen in this world. While the Nibelungen ultimately unleashed magical powers to the bearers of their ring, the physicists are revealing, for the first time, hitherto unseen, mysterious, and ultimate powers, the forces of nature, forces that have sculpted the entire universe in all things from galaxies to stars, humans to DNA, atoms to quarks.

The ring near Geneva is nothing less than the most powerful particle accelerator on Planet Earth. It is called the Large Hadron Collider, or LHC. It belongs to CERN, the *Conseil Européen pour la Recherche Nucléaire* (European Council for Nuclear Research). CERN belongs to Europe and is the largest physics laboratory on Earth, conducting basic research into the inner structure and behavior of matter. Produced in the LHC are the most energetic collisions between subatomic particles ever achieved by humans in a laboratory. In its simplest terms, the LHC is the *world's most powerful microscope*, and it is now the foundry of the "Higgs boson," whimsically known as the "God Particle."

The entire undertaking of constructing and operating the LHC and harvesting its new form of matter, with its far-reaching results for the science of physics and the human understanding of the laws of nature, will ultimately bestow enormous economic power and prestige to its bearers, the countries of Europe that heroically toiled to build and who now operate the powerful ring. And to an extent, the US will benefit for contributing some funding and collaborating in the effort.

But the great LHC and its modern and grand scientific triumph for Europe exists in large measure because the US screwed up big time.

IRONY IN THE 90S

The US botched an even greater undertaking to build an even bigger ring, deep in the heart of Texas, in a picturesque small farm town called Waxahachie, about forty miles south of Dallas, some twenty years ago.

The story of the Superconducting Super Collider (SSC) is a long and winding tale, fraught with eager anticipation; then missteps and setbacks; abounding with technical, political, and fiscal difficulties; and in the end leaving dashed hopes and careers behind. It pains us to think about it all now. The SSC was a grand and noble undertaking, and it would have been the ultimate jewel in the crown of American—no, worldwide—scientific and industrial prowess, a reassertion of that great Yankee can-do know-how of yore. But alas, the SSC died a stillborn death in 1993.

It is the height of irony that at the very same time the SSC perished, several major revolutions were occurring in other spheres that had a common genealogy, all deriving from and driven by basic, non-directed research in modern science. The community of scholars, known as the *academic economists*, who hang out at such places as MIT, The University of Chicago, Princeton, and other vaulted ivory towers, were finally comprehending, in detail, what it is that actually causes economies to grow and prosper.

Astonishingly, over the 200 years since Adam Smith wrote *The Wealth of Nations*, the simple question "What makes economies grow?" had not been answered. Why had the seething misery of the 1820s Dickensian London given way to the hustling, bustling world center of prosperity in the Victorian London of the 1890s? Even the first modern Nobel Prize–winning economist, the great Paul Samuelson—with whose world-renowned textbook many of us slugged out Economics 101—had predicted that after World War II the Great Depression would return. But it didn't. Why not? The opposite happened, as we entered a time of growth and prosperity that lasted to the end of the twentieth century. Why?

Using a modern mathematical theory developed in the 1950s by the Nobel economist Robert Solow, it became possible to calibrate the spectacular growth of the global post–World War II economy. It was found that the spectacular growth was not due to the usual economic activity of bank lending and gambling on commodities futures. There was something else most definitely driving the boom. Some sort of "exogenous input," as Solow called it, was driving the creation of new businesses and new high-quality jobs aplenty. In fact, using Solow's sophisticated mathematical economic model, one could calculate that 80 percent of the postwar growth was coming from this mysterious exogenous input. But what exactly was the exogenous input?

The answer came in the 1990s, just as the SSC was being terminated, largely by the efforts a young and somewhat maverick member of the priesthood of economists named Paul Romer. The answer is almost obvious, yet it took more than 200 years from Adam Smith's *The Wealth of Nations* to figure it out. The answer is (drumroll): *economies grow because of investment in science!* Basic science, applied science, all science. All scientific research pays a handsome dividend, and the more science the better. One should invest in all sciences at once, from green science to hard cutting-steel science, from biology to physics. One should invest in a diversified portfolio. If you want to have a great economy, with jobs and prosperity for all, then you *must* spend your money on basic science. In fact, there is virtually no limit to the return on your investment. And, there is virtually no other way to do it. If you must practice austerity, whatever you do, don't cut the science budgets. And if you really spend enough money on science, you won't need to have austerity![1]

The fact that science drives economic growth is almost obvious to most people (certainly obvious to physicists and their neighbors), yet it took the eggheads of economics more than 200 years to figure it out on their terms. This is a very good thing because it provides a firm foundation to the relationship of science to society and human activities, which in turn provides a basis for making public policy for the investment in science by governments and for incentives to collaborate. Whatever one may think of the actual "science of economics," we believe that Solow's and Romer's (and others') discovery—that economic growth is driven by science—is right and true. When we think of all the many times our colleagues have heroically boarded flights to Washington, DC, to go argue for more science spending with their congressmen, only to return exhausted and disillusioned, we wonder if perhaps they should have also visited the Federal Reserve, where they may have found more sympathetic ears.

THE BIGGEST "EXOGENOUS INPUT" EVER

It's not hard to see the "exogenous input" of science into our economy at work. In fact, and again in the ironic 1990s, perhaps the grandest example of all was playing out. In 1989, a young, obscure, and virtually unknown computer scientist at CERN, Tim Berners-Lee, wrote a project proposal to the laboratory's Computing Division, at which he was employed. Berners-Lee proposed to develop a "distributed information system." Now what, you may ask, is a "distributed information system"? Is it what you get when you throw your unbound PhD dissertation up in the air on a windy day? Even Berners-Lee's boss may have been similarly confused and might have written on the cover of the proposal, "Vague, but exciting," as he gave the green light to the project. Little did he know that he would have just unleashed the greatest information revolution humanity has ever seen and which today garners many trillions of dollars worth of new gross domestic product per year for all the people on Planet Earth.

Tim Berners-Lee conceived of the basic tools that could meet the demand for information sharing over computer networks, initially only between particle physicists, all over the world. He founded the World Wide Web, which has now expanded far beyond the geekish community of particle physicists. It has changed the way we all live, work, and even think. By Christmas of 1990, Berners-Lee and associates had defined the Web's basic concepts, those funny names like "URL,""http," and "html" (never have such cryptic acronyms been typed by so many in such little time and on such a grand scale). They had written the first "browser" and stuff called "server software." The World Wide Web was soon up and running.[2]

In 1991, an early Web system was running for the particle physics community for which it was originally developed. It rapidly began to spread through the academic world of the particle physicists to Fermilab, to the Stanford Linear Accelerator, to Brookhaven National Lab, to the University of Illinois, and beyond, as a wide range of universities and research laboratories started to use it. In 1993, the National Center for Supercomputing Applications (NCSA) at the University of Illinois released its Mosaic "browser," the first modern window-style navigator for the Web that could display in-line pictures and that was easy to install and run on ordinary PCs and Macintosh® computers. A steady trickle of new "websites" soon became a torrential flood.

The world's First International Conference on the World-Wide Web was held at CERN in May 1994 and was hailed as the "Woodstock of the Web." And although Al Gore took some heat for claiming to have "invented the Internet," he did sponsor the key legislation, passed in 1991, that made the ARPANET a high-speed data transmission network, open to the general public.[3] This largely stimulated the usage of the Web and subsequent development of the browsers, as well as new software languages for the Internet, and catapulted the accessibility and ease of use of the Internet for everyone. Soon there would be Yahoo!®, Google®, Amazon®, and countless businesses and exploits and things to browse on the Web, and vast web-based commercial activity, from finding a mate to buying a house or ordering the best coffee and doughnuts. The Web is now blended into the entire worldwide telecommunications system. The economic valuation and impact of the World Wide Web is inestimable.

The Internet and World Wide Web were direct consequences of basic research in the science of particle physics. Particle physics is a worldwide science involving large teams of many people collaborating on single projects, and it was in dire need of a worldwide information-sharing system. It provided the unique and essential paradigm for the development of the World Wide Web. If US particle physics received a mere 0.01 percent (a hundredth of a penny on the dollar) of the tax revenue per year on the cash flow it has generated by inventing the World Wide Web, the Superconducting Super Collider would have been built in Waxahachie, it would have discovered the Higgs boson ten years ago, and we'd now be well on to the next machines—electron colliders, very large proton colliders, and a veritable star-ship of a particle accelerator called the Muon Collider (which we'll discuss later).

THE ROLE OF LEADERSHIP: THE US CONGRESS

We call them our elected "leaders," but it was ultimately Congress that could not seem to find the "leadership cojones" needed at the critical moment in the SSC debacle. In a typical fit of budget austerity, concealed by faint praise for the great scientific endeavor, Congress officially killed the SSC on October 31, 1993, following a key vote in the House of Representatives on October 19. Austerity had become the modern political tool, and American science slowly began to be strangled by it. The new economic theory, indeed the obvious fact of growth driven by scientific research, got, and still gets, no traction on the floor of the US House of Representatives.

Upon perusing the 103rd Congressional Record from 1993 (HR8213-24), we find some of the ironic testimony and the prevailing ill winds of that time. We have provided a *fictional caricature* here to illustrate the gist of it all—any resemblance to real testimony or persons is merely shockingly accidental:

Hon. Mr. X:

"Mr. Speaker, I am afraid that I am stirred into reluctant opposition of any further funding for the Superconducting Super Collider (SSC). The Superconducting Super Collider would indeed be the largest and highest-energy particle accelerator in the world, and it may indeed be the largest scientific facility and allow for the largest and most profound physics experiments ever done. It would even ensure America's lead in science and technology and innovation well into the next century and beyond. Indeed, it would stimulate our best and brightest youth to consider science as a career and to develop a sustainable future for all of us. This unique research tool could perhaps unlock some of nature's greatest mysteries and give us a better understanding of our entire universe. Who knows what new inventions and spin-off products it may lead to, and what brilliant young minds will be inspired and how it will vastly improve our lagging science education system? If we're ever going to have a *Starship Enterprise* in the future, we'll certainly need a Super Collider today. The project would also attract some of the best and brightest physicists and scientists from all the nations around the world to the United States, compensating for our pathetically meager funding for scientific education here at home, further leading to the development of new critical technologies, and securing America's leadership position in fundamental physics research for the next century. Why, it would even help promote peace on Earth!

However, with the mounting federal budget deficit (not to mention severe pressure from my majority whip), it has become increasingly difficult over the past couple of years for fiscal conservatives such as myself to vote for any expensive scientific endeavor. We'll support better technologies for oil drilling, but a Super Collider is just too hard for me to support. (I really haven't a clue what one is.) Federal spending has to be prioritized, and this has been steadily eroding my support for the SSC. We have to get on with starving the beast of a federal government that has such a voracious appetite.

Voting against the Superconducting Super Collider is, for me, a very difficult decision. Much of the SSC's research and development is being conducted in colleges and universities in my own district. Nevertheless, I feel strongly that this is a prudent and responsible vote. If other nations will benefit from discoveries and technology produced by the SSC, why

> shouldn't they contribute to its construction? Or build their own? The Superconducting Super Collider is simply too expensive for the United States to build on our own. And I will not raise taxes on my overburdened wealthy constituents. The economy can only grow by the efforts of American businessmen. In fact, I'm planning to sign a brand new "pledge," crafted by my dear friend, Mr. Grover Norquist, that I will never vote to raise taxes again.
>
> Our country has been a world leader in technology, and we must continue to support our nation's scientific research programs. The SSC, however, is a program that we simply cannot responsibly fund. Although proponents have argued that the potential scientific benefits outweigh the high costs and that the SSC should be an immediate priority, I'd just like to see a few more roads in my district and a lot more oil drilling and coal mining. And, I might add that if humans were meant to see quarks and "God Particles," God would have given them tiny eyeballs.

The final vote to support the continuation of funding for the SSC in the House of Representatives was 159–264, the nays having their way with it. The construction of the tunnel for the Superconducting Super Collider was about a third complete, and more than two billion dollars had already been spent.

It wasn't just Congress's fault. There were many problems in the management and attempted execution of the SSC project and plenty of blame to go around. It's very hard to pinpoint exactly what killed the SSC because it was a mélange of reasons. We don't want to indulge in dredging it all up, and the interested reader can read about it at any of the numerous websites provided to you by high-energy physics and Mr. Tim Berners-Lee and associates.[4]

But for science, all of science, and as an engine of economic growth and prosperity in the US, the termination of the Superconducting Super Collider was an unmitigated disaster. If Congress was truly a body of *leaders*, it would have moved forward on the SSC and found a way to make it happen.

The massive drilling machines of the SSC were abandoned in the half-completed tunnel where they stood. The tunnel itself carved in the frail Austin chalk, its sump pumps unplugged, has now filled with water, its walls soggy and collapsing, the heavy equipment dissolving away like some far-off shipwreck. Sage brush once again rolls in the wind through the downtown streets of Waxahachie, as plywood panels creak and shutters clap against walls, like some late-night replay of *The Last Picture Show*.

IS IT RECOVERABLE?

Over the years big science has increasingly fallen victim to the political process. Sadly, in the US, it seems not to be recovering. Particle physics in the US has not enjoyed the construction of a new cutting-edge particle accelerator since Fermilab's Tevatron was built in the 1980s (expanded with its Main Injector in the early 1990s). Big science in other fields, such as nuclear fusion and astrophysics, has also lost many similarly large-scale projects. Today, many people now question whether modern American-style democracy can ever manage grand science endeavors again. They wonder, are Americans, through visionless leadership, endless partisan bickering, and Machiavellian ultra-rich special interests and their lobbyists now destined only to sit watching TV in their devalued homes, fearing the mortgage collector, their jobs threatened, with their devalued currency, fighting multifront wars, making a mess of the planet with their fossil fuels, while the grand discoveries and exploration of the ultimate frontiers—and the burgeoning future economies this will bring—happen in Europe, China, India, and elsewhere? With the discovery of new phenomena, such as the Higgs boson at CERN in Geneva, Switzerland, will indeed come direct and indirect applications that can resuscitate moribund economies. But is this is now less likely to happen in America?

Fortunately, at least for American scientists, the US does significantly participate in the big experiment collaborations that do their physics at the LHC, and the US significantly contributes technology and manpower to build it. Particle physics has grown up to become a truly international activity, and no large particle accelerator will likely ever be built again without full international collaboration. Even the use of the Fermilab Tevatron, which was the leading particle accelerator for two decades prior to the LHC, was an international collaborative effort, and Fermilab, like CERN, flies the colorful flags of its collaborators' nations in front of the main building.

But there's no place like home, and not having such a large-scale project on American soil makes the economic gain diffuse and less powerful in the long run. Schoolchildren in the Chicago area can now only visit a museum showcasing a past great particle accelerator at Fermilab—not an operating facility collecting never-before-seen data about the inner universe of matter.

Yet, as regards the future of Fermilab and the US particle physics program, we're eternally hopeful. We'll discuss what we *could* and *should* plan to do to change this and move forward. There is a pathway back for American science, but we'll certainly have to refresh our American aspiration for greatness and find our old our "get up and go" that seems to have "got up and went."

LONG LIVE THE KING

The LHC is operating superbly today. It is a particle accelerator that hurls protons at enormous energies head-on into one another, at the highest energies ever achieved by humans. Within every trillion collisions of the protons at the LHC there emerges a mysterious new form of matter—shards of matter that exist for only a billionth of a trillionth of a second—yet enough time to be recorded in the two great eyeballs of the LHC, the particle detectors known as ATLAS and CMS. The market value of this form of matter cannot be assessed—it's meaningless—its value is determined by the cost of the entire project of the LHC and, per ounce, it is literally trillions of trillions of trillions times greater than that of gold.

With the demise of the SSC, the US walked off the playing field of the highest energy particle accelerators. The US has essentially outsourced this, perhaps the most important science, to someone else. And that "someone else" is CERN, and the Europeans know what they are doing and they are doing it well.

So, what is this place called CERN?

Western science began in Europe. It began with the ancient Greeks, but the modern era traces to Galileo, who divined the law of inertia by studying balls moving on inclined planes. Newton encoded this into his laws of nature and discovered that gravitation is a universal force that permeates the universe, holds the earth in its orbit about the sun, and controls the fall of an apple in a garden.

This was the birth of the Age of Enlightenment. It ultimately led to the summation of the laws of electricity and light by Faraday and Maxwell, to Max Planck and Albert Einstein and the twentieth century leap forward with the discovery of the quantum behavior of all small things. Countless phenomena that were opaque and mysterious could now be attacked scientifically. The atom, the chemical bond, and the chemical basis of life as well as the properties of materials and properties of the elementary building blocks of all matter could now be understood. But the political climate in Europe shifted toward horror, with the rise of fascism in the twentieth century, and many of the greatest European scientists had to leave, including Einstein, Fermi, Emmy Noether, and many others.

By the end of the Second World War, European science had lost its leadership role held for the three and half centuries since Galileo to the United States of America. However, a small group of the leading European scientists, including most notably Niels Bohr of Denmark and Louis de Broglie of France, envisioned creating a new center for physics in Europe. Such a laboratory would stimulate European scientific research but would also permit sharing the increasing cost burdens of the large-scale facilities required for nuclear and particle physics.

> French physicist Louis de Broglie (one of the founding fathers of the quantum theory) put the first official proposal for the creation of a European laboratory forward at the European Cultural Conference in Lausanne in December 1949. A further push came at the fifth UNESCO General Conference, held in Florence in June 1950, where the American Nobel laureate physicist, Isidor Rabi, tabled a resolution authorizing UNESCO to "assist and encourage the formation of regional research laboratories in order to increase international scientific collaboration . . ." In 1952, 11 countries signed an agreement establishing a provisional Council—the acronym "CERN" was born and Geneva was chosen as the site of the future Laboratory. The CERN Convention, established in July 1953, was gradually ratified by the 12 founding Member States: Belgium, Denmark, France, the Federal Republic of Germany, Greece, Italy, the Netherlands, Norway, Sweden, Switzerland, the United Kingdom, and Yugoslavia. On 29 September 1954, following ratification by France and Germany, CERN officially came into being.[5]

In 1957 CERN built its first particle accelerator, a comparatively low-energy machine that provided particle beams for CERN's first experiments. It evolved into a machine that was used for research in nuclear physics, astrophysics, and medical physics, and was finally closed in 1990, after 33 years of service. The original synchrotron had given way to a more powerful "Proton Synchrotron" (PS) by late 1959, which still operates today.

Conventional particle physics experiments are identical in configuration to those of a biologist's microscope—a point we'll be harping on throughout this book. Think of a microscope in a typical high school biology lab. Here you have an incoming light source (the particle beam) colliding with some material that you've placed on a glass slide, perhaps containing a drop of pond water (the target), in which there might be a paramecium or an amoeba swimming around (the "quarks" you want to see). The incoming light scatters off the target and is collected in an optical system with lenses that magnify the image and present it to your eyeball (the detector). In this way you can see the little microbes swimming around in their drop (data!). So, in summary we have (1) a particle beam; (2) a target; and (3) a detector that collects (4) data. That's it—very simple indeed—these powerful accelerators and their detectors are microscopes.

There's one key point, however, about particle physics that you must grasp: the smaller the thing you want to see in the target, the higher the energy you must impart to your beam particles. The reasons for this will be explained later, but it is our basic operating principle. This is also true for microscopes, and it's why electron microscopes, which use higher-energy beams of electrons instead of low-energy beams of visible light, "photons" are better than optical microscopes. This is actually where microscopes really do become particle accelerators—it's more than just a powerful metaphor because it's really true, and all the issues and challenges of building powerful microscopes hold as well for particle accelerators, and vice versa.

In the usual way of doing their particle collision business, physicists make a particle beam strike an atom that is sitting at rest in a fixed target, like a block of lead or beryllium, just like the light beam striking the target on the glass slide under the microscope. However, physicists realized that in the subsequent particle collisions with the atoms in the block of material, much of the precious energy of acceleration of the incoming beam is wasted. The outgoing particles emerging from the collision acquire "recoil momentum," which takes away the useful energy. However, if two particle beams could be fired at each other and made to collide head-on, then there need be no recoil momentum and all the energy is available to probe deep inside of matter. The full particle beam energy becomes available to make a detailed image, or to actually produce new and previously unseen and very short-lived elementary particles. This is the concept of the modern "particle collider." The Fermilab Tevatron and the CERN LHC (previously CERN's LEP collider) were and are all-powerful and useful particle colliders. But what a challenge this is—to make infinitesimally small particles that are smaller than a millionth of a billionth of a golf ball in size, traveling almost at the speed of light, hit each other head-on!

The world's first proton–proton collider, called the Intersecting Storage Rings (ISR), was built and came into operation at CERN in 1971. The ISR was a very small machine by today's standards, but there were daunting challenges to overcome just to make it work. The ISR produced the world's first head-on proton–proton collisions. It was actually constructed on French soil on land adjoining the original CERN site in Switzerland. At the same time, the first electron–positron collider was ramping up at Stanford Linear Collider laboratory.

THE GRAND SYNTHESIS

By the early 1970s theoretical physicists had sewn all the available data from a century of research together, much of it coming from the data produced at high-energy particle accelerators, and had developed a remarkable descriptive and predictive theory that has come to be known as the "Standard Model." One of the great achievements of the Standard Model is that it united two of the known forces in nature into one unified entity. These two forces are *electromagnetism*, the force associated with ordinary electricity, light, and magnetism, together with a very feeble force, so feeble that it wasn't even noticed until the 1890s, called the *weak interactions*. Though this latter force is "weak," without it the sun could not shine and we would not exist.

The electromagnetic force is associated with particles, called *photons*, that are the particles of light. Likewise, the Standard Model predicted that the weak force must be associated with three previously unseen particles, called the W^+, W^-, and Z^0 (these are called the "weak bosons"; "bosons" are defined in the Appendix). These three particles were predicted to be very heavy by particle physics standards (the W^+ and W^- are 80 times heavier than the proton, and the Z^0 is 90 times heavier than the proton), and they have extremely short lifetimes, less than one trillionth of a trillionth of a second. But physicists realized that these particles could in principle be produced and detected in a sufficiently energetic collider experiment. Alas, no machines existed at the time the Standard Model was put together that were capable of producing the weak bosons, but indirect hints of their existence continued to emerge in various experiments. These "indirect" hints compelled the ultimate construction of a machine capable of directly producing and observing the W^+, W^-, Z^0.

In the early 1970s the first big accelerator at Fermilab (called the Main Ring) came into existence, while CERN had built, upon their existing PS, the Super Proton Synchrotron (SPS). Both the Fermilab Main Ring and the CERN SPS were a whopping four miles in circumference (approximately). Particle physics had become big science. These were powerful accelerators that could be used to produce very energetic beams that went off to various fixed target experiments. With fancy upgrades, however, they could in principle be converted to colliders, and with head-on collisions they could produce and "discover" the W^+, W^-, Z^0.

In the late 1970s Fermilab embarked upon an ambitious and long-term goal of building the Tevatron, a machine that would collide protons with antiprotons[6] and that would ultimately become the world's first superconducting-magnet collider operating at the highest achievable energies for the four-mile circumference ring. CERN, on the other hand, took the bold decision to convert the SPS into a proton–antiproton collider to aggressively stalk the weak bosons, W^+, W^-, Z^0, as quickly as possible. This was a risky gambit, but it paid off handsomely.

The first proton–antiproton collisions at the CERN SPS were achieved just two years after the project was approved. Two large experiments at the SPS, named UA1 and UA2, started to search the collision debris for signs of weak interaction particles, and in 1983, CERN announced its discovery of the W^+, W^-, and Z^0 bosons. Carlo Rubbia and Simon van der Meer, the two key scientists behind the discovery and the SPS conversion into a collider, received the Nobel Prize in Physics within a year. Fermilab's Tevatron came online later. It should be noted, however, that throughout this period the Fermilab budget (in today's dollars) was about $300 million per year, while the CERN budget was more than $1 billion per year. Money may not buy happiness, but it does buy big science, quickly and effectively.

Though scooped by the W^+, W^-, Z^0 boson discoveries at CERN, the two Tevatron experimental collaborations, known as D-Zero and CDF, successfully discovered the elusive top quark, the heaviest of all known Standard Model particles, in the mid 1990s. All that remained to find in the Standard Model was its missing link—the Higgs boson. As the Tevatron assumed the role of the world's highest-energy particle accelerator, CERN began to build a new kind of collider, and the physically largest one ever. The new machine was a collider of electrons and their antiparticles, positrons. This was called the "Large Electron–Positron Collider (LEP).[7]

It was thought that LEP might actually discover the Higgs boson, an ingredient of the Standard Model that had been hypothesized by theorist Steven Weinberg to provide the origin of the masses of all the particles. The optimism of a LEP discovery had sprung from certain popular theories that had argued the Higgs mass was actually less than that of the Z^0 boson. To achieve the required energies to make a Z^0 boson with the precision afforded by using electrons and positrons (see chapters 7 and 8), LEP had to be an enormous circular ring, housed underground in a deep tunnel. CERN therefore built a 27-kilometer (almost 17 miles) circumference circular tunnel, the construction of which ultimately proved decisive for a pathway to the LHC.

The excavation of the LEP tunnel was Europe's largest civil-engineering project prior to the Channel Tunnel. The tunnel is actually tilted, with its high side under the Jura Mountains, and this presented enormous and somewhat unforeseen engineering challenges, particularly in controlling high-pressure water leaks from springs within the mountains that threatened nonstop and major flooding. Three tunnel-boring machines started excavating the tunnel in February 1985, and the ring was completed three years later.

LEP did not find the quarry it had sought—the elusive Higgs boson. In a machine such as LEP the "signal-to-background" ratio for the Higgs boson would have been optimal for a discovery. Though many scientists expected the Higgs boson to be within LEP's reach and were disappointed at the Higgs boson's failure to emerge at LEP, the machine nevertheless made remarkably detailed precision measurements of the properties of the Z^0 boson that have had major impact on our detailed understanding of the Standard Model. LEP was even upgraded for a second operation phase, to produce and study W particles, but the energy of the machine still limited the reach for a Higgs boson to a mass scale of less than about 115 times that of the proton. It still did not, even with the higher energies, find the Higgs boson. In the meantime, the SSC was canceled in the US, and CERN saw a golden opportunity to convert the LEP collider to a much higher-energy proton–proton collider. The LHC project was born.

The LEP collider was closed down in November 2000 to make way for the construction of the Large Hadron Collider (LHC) within the same tunnel. The LHC was built and commissioned, and it is, today, the world's most powerful particle accelerator, the most deeply penetrating tool we have into the inner workings of matter and energy.

COMMISSIONING THE WORLD'S LARGEST PARTICLE COLLIDER ISN'T EASY

Einstein said you should always drill through the thick part of the wood. Progress can only be made by extraordinary effort. Extraordinary effort often implies overcoming extraordinary setbacks. Particle physics drills deeper into the wood than any other human endeavor. It must necessarily have major setbacks from time to time.

The Large Hadron Collider at CERN spans the border between Switzerland and France, sitting about 300 feet underground in the LEP tunnel, a large circle with a circumference of about 17 miles. The LHC is the world's most powerful microscope and is used by physicists to study the smallest known particles and processes—the fundamental building blocks of all things. Two beams of protons travel in opposite directions inside the circular accelerator, gaining energy with every lap. Particles from the two beams collide head-on at very high energy within the centers of the two enormous detectors ("eyepieces") called ATLAS and CMS. Teams of physicists from around the world then analyze the collisions. Two additional medium-size experiments, ALICE and LHCb, have specialized detectors that analyze the LHC collisions for a wide assortment of other phenomena.[8]

Fermilab, with many US universities, collaborates mainly with CMS, and many US university physicists work with ATLAS. Fermilab built the LHC Remote Operations Center for the CMS experiment within its main building, Wilson Hall. Here US scientists can be part of the action and play a vital role in actually managing the operation of the experiment without having to hop on a transatlantic flight.

On September 10, 2008, 1:30 a.m. in Fermilab's Remote Operations Center, the first circulating proton beam in the LHC in Geneva was celebrated by a partying audience of dozens of scientists, who were pulling an all-nighter, many clad in robes and pajamas. The mood was confident and exuberant. A new age for particle physics was dawning.

As the particle energy in the LHC is increased, the magnetic field strength in the accelerator beam pipe must also increase to hold the protons in fixed circular orbits within the machine. This is accomplished with 1,232 special "dipole magnets" (and thousands of other magnets that serve effectively as "correcting and focusing lenses" in the system; the system has about 5,000 magnets in total[9]). These magnets use electrical current to generate controlled magnetic fields. This can be varied to sweep over the required field strengths needed to maintain the delicate particle orbits, as their energies increase during acceleration (this process is called the "ramp"). Each LHC dipole magnet is a massive steel structure, about 50 feet in length. Their magnetic field ranges from zero to 100,000 times that of the earth's magnetic field. When the maximum field strength is reached, each magnet contains the potential energy of about one-quarter of a ton of TNT. This energy would be accidentally and explosively released if the current flow in the magnet were instantaneously disrupted.

To operate efficiently, the magnets use the principle of superconductivity and must be cooled down to near absolute zero temperature (0° Kelvin, which is 459° Fahrenheit). The magnets are housed inside enormous cryogenic containers, called "cryostats," which contain the ultra-cold liquid helium. At temperatures approaching absolute zero the special magnet coils, made of copper-clad niobium-titanium cables, become superconducting, and then offer no resistance whatsoever to the flow of electrical current. Without this, the power bill to operate a collider would be unaffordable.

For the days after the initial beam was circulated, and the Fermilab pajama party on September 10, the accelerator physicists at CERN began to gradually step the machine up to its higher design energy. The precision-engineered coil windings of these magnets must be secure against any tiny movements as the magnetic field becomes stronger. This ultra-strong magnetic field exerts enormous stress on the magnet structure, and the structure must literally contain the force of the field against an explosion. Slight changes in the magnet can create "normal," or non-superconducting "hot-spots." These hot-spots could "quench" the magnet, such that it loses its superconductivity, by being driven out of its cold, superconducting state.[10]

In a quench, to prevent an explosion, the enormous electrical current flowing through the magnet coils must be quickly drained out of the system. A "quench protection system" is in place to minimize the effects of any unwanted quench incident. Astonishingly, copper, which is a good conductor of electricity at normal temperatures and is used in the wires of your home, is actually used as an *electrical insulator* on the superconducting cables, since the current will always flow through the superconductor and not the copper at the low temperatures! If a hot spot quench arises, the current can then be safely carried away through the surrounding copper, minimizing any damage to the system. A quench in any one of the totality of about 5,000 LHC superconducting magnets could disrupt the machine operation for several days with the quench protection system, but would be catastrophic without it.

Superconducting magnets have to be "trained" to reach higher and higher magnetic fields, as smaller and smaller glitches are relaxed from the coils. The engineers use advanced computer monitoring systems to watch for any possible quenches and induced stresses before they develop into larger problems. The enormous currents that flow within any particular magnet must also pass onto the next magnet. This is done outside of the superconducting environment of the cryostats and requires enormous copper junctions, joined together at face-to-face soldered copper plates. Even these low-tech solder joints are monitored by computers for any changes in temperature during the operation of the system.

OH, $%&#!

By September 19, 2008, things had settled back to "business as usual" at CERN. The LHC was being ramped up to full magnetic field strength as the magnets were being trained. Gradually, carefully, and systematically the LHC operators in their Swiss control room, with the precision of watchmakers, pumped more and more electrical current into the massive superconducting magnets that steer the beam in its 8-kilometer (5-mile) diameter (that's the same as a 27-kilometer, or 17-mile, *circumference*) circle.

Suddenly . . . *a massive cataclysmic explosion ripped through the tunnel!*

The electrical circuit feeding one magnet to an adjacent one had inexplicably "opened." Later it would be discovered that the solder joint between external copper plate connectors conducting the electrical current from magnet to magnet had melted. While the monitors had detected some heating in the copper joints, evidently the meltdown of the solder happened too quickly to be detected. The soldered joints had failed and the electrical circuit opened up.[11]

When the current flow through an electromagnet is suddenly interrupted, i.e., when the circuit is "opened," there develops a virtually infinite voltage. This happens because nature opposes the interruption and strives to restore the current flow that is required to maintain the magnetic field. This is the basis of spark coils in automobile engines, Tesla coils, and any devices that step up voltage from one value to another, such as a transformer. For the largest magnets in the world, opening of the circuit produced a monstrous spark. At the LHC this human-made bolt of lightning, like a super-bolt from Odin's scepter, blasted through the neighboring magnets and pierced the cryostat, the large vessel that holds all of the liquid helium that keeps the magnet cold and superconducting.

The result was an explosive release of 3,000 gallons of liquid helium into the confines of the LHC tunnel. The resulting shock wave traveled down the tunnel at the speed of sound. A mile away steel doors were blown off their hinges. At least five of the massive magnets were destroyed, and about fifty others were damaged. The accelerator's entire vacuum tube was corrupted, as debris was sucked back into the beam pipe around its entire 27-kilometer circumference. It is said that the oxygen and nitrogen within the vicinity of the explosion was condensed out of the air itself and lay on the floor of the tunnel in a pool for six hours.

As a testimonial to modern safety standards, no human being was even slightly injured in this monstrous explosion. The catastrophe occurred at the time of international economic turmoil, however, and threatened the viability of the LHC project. The entire accelerator had to be cleaned out, and major repairs were required along nearly a half mile of the tunnel, including the replacement of five magnets. The event revealed a design flaw with the non-superconducting copper joiners that connect one magnet to another, maintaining the current flow through the system, and these had to be replaced around the entire 27-kilometer circumference of the machine.

Though the entire future of the LHC was in doubt for a brief time, CERN persevered and brought it back online within two years. This was a spectacular and heroic feat, the kind of challenge that any risky new and cutting-edge endeavor must overcome, reminiscent of NASA's near-disaster with the Apollo 13 lunar mission. The LHC came back online, and less than four years after the "helium incident" successfully discovered the Higgs boson. LHC's performance since then has been jaw-droppingly spectacular. CERN's achievement with the LHC would surely have put a smile on the face of Albert Einstein—it has certainly drilled through the thick part of the wood.

At this writing (March 2013), the LHC is down for upgrading and will come back online around January 2015 for what will be the most important survey of the highest energy scales, or shortest distance scales humans have ever probed, at energies well beyond the mass of the Higgs boson.

So, since the LHC is the world's most powerful microscope, what is it we are looking for?

"ACH! IF I'D KNOWN THERE WOULD BE SO MANY PARTICLES I WOULD HAVE BEEN A BOTANIST INSTEAD."

So said Einstein, supposedly, tongue in cheek, to a colleague in the lunch line at Princeton, who was explaining to the great master something of some newly discovered particles in the early 1950s. Back then these were called the "strange" particles, and they are known today to be composed of things dubbed "strange quarks." The understanding of the nuclear particles, those associated with the strongly interacting particles found in the atomic nucleus, spanned the 1950s well into the 1970s. This branch of physics is still going on today in a modern form as we contemplate the ferociously energetic collisions at the LHC. Today, in the dawn of the third millennium, what is the new form of matter we would like to tell Professor Einstein about? That is the multi-billion-dollar question—the subject of this book: it's the Higgs boson, or the God Particle. And, as to what lies beyond—we simply don't know.

So, we might tell Professor Einstein in the lunch line at Princeton about the Higgs boson, and it goes as follows:

In its simplest theoretical incarnation, the Higgs boson is a form of matter named after one of the physicists who first considered the possibility. It forms a field, something like a magnetic field that is composed of photons, that fills all of space. Particles of matter interact with this field and acquire their masses. Heavy particles, like the top quark, interact more strongly with the Higgs field than light ones, like the electron. This should more properly be called the Englert-Brout-Higgs-Guralnik-Hagen-Kibble Mechanism after all of the authors who should share credit for it.[12] The idea of the Higgs boson, and the way in which it gives mass to other particles in nature derives from many sources in other fields of physics as well. The central idea, in fact, lies at the heart of superconductors and was first considered by people like Fritz London in the 1930s.[13] Following the initially somewhat general ideas of the Englert-Brout-Higgs-Guralnik-Hagen-Kibble Mechanism, Steven Weinberg put it all together and showed precisely how such a particle would fit into the overall scheme of nature, which ties together the so-called "weak" interactions with the "electromagnetic" interactions to form what we now call the Standard Model.[14] The name "Higgs boson" stuck. The Higgs boson does the job of making all the particles have their masses, and it becomes an essential ingredient. Through the Standard Model, the Higgs boson properties become precise, though the Higgs boson mass itself is *ab initio* unknown. Given knowledge of the Higgs boson mass, the Standard Model tells us how to produce the Higgs boson and how it will show up in detectors through its telltale fingerprints (i.e., via its "decay modes"[15]).

That's the "sound byte" (more of a "sound paragraph") we might tell Professor Einstein, but throughout the next few chapters it is our goal to explain in clear and simple terms what this means, yet in greater detail, to give you a feeling as to why we need the Higgs boson and to give some inkling as to what may lie beyond.

WHAT'S IN A NAME?

By 1993 the search was already well under way when the Higgs boson was poetically dubbed the "God Particle" by Nobel Laureate Leon Lederman (one of the authors of this book, if you haven't noticed) and his co-author, Dick Teresi.[16] This is the title of Leon's entertaining and poignant autobiography, which can still be purchased on the World Wide Web at Amazon as a bargain paperback or downloaded for your Kindle.

Leon Lederman, aside from having made several major discoveries in physics that gave him the Nobel Prize in 1988, is also known for his talent in telling jokes (*The Lederman Show* could just as well have been hosted by this Lederman). The term "God Particle" was a tongue-in-cheek "handle" for the Higgs boson, and it caught on, soon appearing in popular journalism covering the forefront developments in the science of particle physics, from the *New York Times*, to *Der Spiegel*, from the *Jerusalem Post* to the *Pakistan Chronicle*.

Indeed, the moniker "God Particle" has stirred people up. While it was only an exercise in literary license, many people think it somehow imbues a deep religious significance to a particular elementary particle. It doesn't. Some scientists are disgusted by it, thinking it compromises the dignity and virtuous rationalism of the scientific community, furthermore corrupting the purity of essence and vaulted alabaster image of the scientist hard at work.

We have experienced the mixed consequences of the "God Particle" moniker in public firsthand. Every few years we host an "Open House" at Fermilab, where people from the community can come to see the accelerators and detectors and talk to scientists about what we do. People drive in from many distant Midwestern states for this event. Shortly after the appearance of the "God Particle," we had such an Open House and the turnout of fundamentally religious people had significantly increased—a wonderful thing, as our tent is big enough for all who wish to come—but we suspect many came who sought "God Particles" and were somewhat disappointed by what they learned about our true scientific view of nature and, especially, its view on the creation of the universe. Here's one account of an incident:

> I was serving as a guide at the Open House at this time, and one woman I met asked me rather suspiciously if there was also a "Satan particle?" "No," I said. She continued, "So, exactly how did the God Particle create the world? I don't see it anywhere in Genesis or anywhere in the Bible."
>
> I explained to her about *mass*, an essential property of matter, and how we see it in the "Standard Model" of particles, and how it has to

> do with something called the Higgs boson, and how the "God Particle" is just a whimsical literary name for that. "Mass?" she said, ". . . as in the Catholic Mass?" "No, I replied, "mass is a measure of how much matter there is in something." She glazed over, and I again told her this was only poetically dubbed "the God Particle" by our venerable colleague, Leon Lederman, in his book, a copy of which she was actually holding.
>
> She seemed reengaged, and asked, "Well, where did all of this come from?" So I went on to explain a little about "creation," as scientists see it, the "big bang," "matter-antimatter asymmetry," the "nucleosynthesis" part of it by which ordinary matter is created in the big bang, etc., and that most of the primordial matter is hydrogen and helium, and it all was here within three minutes—that's before a considerable amount of processing that subsequently happened in stars to make the heavier elements, the stuff we're made of. But, I explained, the raw materials were established in the big bang, and many mysteries abound, like "where did the antimatter go?"
>
> She seemed engaged in this and she asked a number of intelligent questions, finally getting to: "But what about life on Earth?" I explained that people weren't around at all for about thirteen and half billion years, and that we ultimately descended from microbial life-forms that arose, well, from large molecules, and eventually evolving into worms, vertebrates, primates, and then, "us."
>
> At this point the woman looked utterly horrified, turned, and hastily beat an exit from the building, returning to a large bus parked outside with the name of a church in Missouri on its side panel. I am sure she felt that while, perhaps, he has no particle of his own, she had surely just met Satan, or one of his many cacodemons, incarnate.

Leon Lederman, unflapped by all of the controversy, sits charmingly and serene as ever in his oversized leather desk chair with a smile on his face and a glowing sense of humor about all of it, the Einstein bobblehead doll on his desk bobbling in approval. The "God Particle" moniker has become a standard by-line in the latest newsworthy updates on the Higgs boson around the world, from Tibet to Timbuktu. "Timbuktu?" says Leon with a twinkle in his eye, "I have an uncle who sells bagels in Timbuktu . . ."

In any case, for the purposes of this book we'll assume that Leon had the iconic Norse god Wotan (Odin) in mind when he named the "God

Particle." And, in fact, the Higgs boson is not the end, or the "ultimate," or even the "Götterdämmerung" of nature and science, but rather it all goes way beyond the Higgs boson. A Higgs boson represents the entry into a new domain of nature, the beginning of a new set of puzzles, and the beginning of our quest to discover something completely and radically new. The story always seems to turn out to be much bigger and grander than we may think it is at any given time.

Indeed, a curious parallel to the Norse myth continues: After its fabrication, Wotan (Odin) donned the Nibelungen's ring and went forth in his earthly wanderings, eventually ceding the ring to Siegfried, who slew dragons and rescued the beautiful Brunhilde from her eternal sleep on the fiery top of a volcano. Ultimately, in Götterdämmerung at the end of the Ring Cycle, the gods perish through their own perfidy and tomfoolery with the golden ring, ceding the future world down to the humans.

The message is clear: we must progress beyond a belief in demi-gods, dwarves, trolls, and selfish and angry gods—beyond the fairy tales we are taught as little children. The world ultimately belongs to and is stewarded, for better or worse, by humans. Perhaps the present moniker "Beyond the God Particle" fits all of this. Humans are continually making progress in learning how the universe really works, beyond fairy tales and myths, and through such profoundly successful international collaborations as at Fermilab and CERN, learning how to work and live together across national boundaries and cultural frontiers. It's all about collaboration on the largest scales of human endeavor. It's ultimately all about the future of people.

And so, we'll now abandon the term "God Particle" and look in greater detail at the Higgs boson, at the science of the smallest things in nature and what we are actually trying to do, and in many ways are succeeding in doing now—what we will achieve with the LHC, and what we hope to do, and must do, in the future. We are also looking beyond the Higgs boson, both as a thing and as an idea. With the Higgs boson in hand, physicists now have a powerful new insight into how nature generates its fundamental patterns and its properties of the elementary particles, and a new, powerful way to understand the remaining mysterious puzzles of the physical world.

CHAPTER 2

A BRIEF HISTORY OF THE BIG QUESTIONS

The most fundamental of questions we are asking today concern the smallest objects, objects that lie far beyond the atom, the quarks, the leptons ("matter") and gauge bosons ("force carriers"), the Higgs boson, and whatever lies beyond these things. Here we are exploring a strange new world—a world of the smallest things. No one has ever been here before, to examine what is happening at the smallest distances that are now probed by the Large Hadron Collider (LHC). This is not entirely blind exploration, for we actually have an inkling of what we are trying to understand—but surprises may be around the next corner.

In short: we are attempting to answer the vexing question: *What is the origin of mass?* Mass is one of the most important defining quantities of matter. But where does it come from? What makes mass happen? Will we ever become skillful enough to calculate the mass of the electron or the muon or the top quark from a "first principle"? What shapes and controls and sculpts the elementary constituents of matter and their masses?

This is a bit like trying to answer the deep biological question "What and where is the genetic code of life?" The answer to that question came in the 1950s—it turned out to be encoded into a very long and durable molecule called DNA. And from that has come an entirely new set of capabilities, as DNA can be "read" and "reread" and, eventually, we think, "rewritten." All structure and function and ultimately all diseases of living organisms are controlled by DNA and its associated processes. Understanding DNA and its evolution is the foundation of understanding all life on Earth. Our open physics questions today are much like the biological ones before the 1950s: "What causes the phenomenon of mass?" Put another way, "What is the DNA of matter itself?"

To get some insight into the process of the exploration of nature, let us ask, what deep questions were our ancient ancestors asking over the past three millennia? Like newborn babies, our ancient ancestors awoke with rational minds and conscious awareness into a world with a "reality" of its own. It was difficult initially for them to shake off primitive prejudices, notions and fears, unwarranted or otherwise, about things that seemed to happen or were only imagined to happen. There was an internal reality to the human mind in the early dawn of intelligence, voices that spoke in the night, apparent demigods lurking behind every tree, making all things, good or bad, happen. This led to peculiar notions, for example, that one must dance in strangely ostentatious ways, while wearing bizarre make-up and costumes, in order to make good things happen, perhaps to make it rain. Indeed, most appeals for divine intervention are just a variation on a rain dance and are motivated by something like the mortal fear of crop failure. It was difficult to discard that and to create a distilled "objective reality."

But gradually there emerged a coherent understanding and philosophy of objective reality. Questions could now be posed and answers sought without reference to mythical beings and magic, without the fear of offending the particular gods that brought the rains. One learned to do "experiments." And one learned that the reproducibility of an experimental result was far more important than the mere opinions of the witch doctors and high priests. Does it really rain when we put on our costumes and dance about? No. But there are certain crops that can grow better in a dry climate than others, and certain clever ways to grow them. At some point the issue of understanding reality became "science."

Eventually people asked the deeper questions: "What are all things made of?" "What are their properties?" "How do they interact with one another?" "What are the fundamental laws of nature that govern these objects?" These are practical questions, but they are also the biggest questions. They deal with profound issues: "What constitutes physical reality?" and "What is the nature of physical substance?" and "What is physical force, motion, space, and time?" The answers hold deep secrets, and perhaps the key to a better fire, a better sword, a cure for illness, perhaps a way to make the rains come or prevent them from leaving, or to make the best of what the conditions are, and how not to mess things up. By the end of the nineteenth century, here on Earth, the question: "What is the nature of matter?" was framed within the province of chemistry: All matter is formed from the basic atoms that comprise the chemical Periodic Table of the Elements—where "periodic'" refers to their chemical properties. The elements form chemical compounds and enter into chemical reactions according to specific empirical rules. The laws of physics are those of Galileo and Newton, embellished by Maxwell, Gibbs, Boltzmann, etc.

Many thinkers from antiquity had previously developed a rudimentary concept of "elements." These would be the basic, irreducible components out of which things are made. Among the earliest ideas were the so-called "classic elements," as described by Plato: "Air," "Fire," "Earth," and "Water," as well as mysterious "Quintessence." The latter was considered to be an all-universe-filling "ether." This view of the nature of matter reduced every question to the five classic elements and offered a (very) tiny hint of an underlying order, but it certainly didn't get into the details. It was more of a dismissive answer to questions about the inner nature of matter.

Other philosophers of antiquity, however, were actually quite modern from our perspective. The foremost of these was *Democritus of Athens*, one of most advanced thinkers in all of human history, considered by some to be the "father of modern science," certainly the Galileo of his age. Democritus was born around 470 BCE, and died around 370 BCE, thus living to the ripe old age of about 100.[1] He was often viewed as an eccentric fellow and largely ignored in his home town of Athens, and was supposedly detested by Plato, who denied ever meeting him (though this was unlikely since Plato allegedly wanted all of Democritus's books burned).

Democritus inherited the moniker "the laughing philosopher," as he evidently found most of the ideas of other contemporary philosophers to be rather humorous, if not ridiculous. We can imagine him heckling Plato during a lecture in some curia, circa 400 BCE, perhaps asking a subtle and detailed question about a certain chemical reaction about which Plato could not begin to answer:

P: And the natural order and simplicity of nature is simply that all things can be resolved to the five "elements," the "air," the "fire," the "water," the "earth," and the "quintessence," and that's all of it.

D: Master, are these elements transmutable into one another?

P: No, truly not, sir, for as I say, they are *elemental.*

D: But of what element is the brilliant light of the sun?

P: *(pause)* I suppose . . . a form of quintessence as it does flow though space which is filled of quintessence and so it must be such.

D: And, master, of what element is papyrus?

P: Surely, papyrus is a form of the earth as it comes from the earth.

D: So, master, if I place a gem of spherically shaped quartz between the position of the sun, and that of a papyrus scroll, which you say is a form of the earth, I can direct, or "focus," the sun-light, a form of quintessence, upon the papyrus and I can produce a fire. Have I not converted the quintessence into the fire or the earth into the fire?

P: I do not believe this can happen, sir.

D: I have set up the experiment here, master *(Democritus directs Plato and the audience to a window at which he has an apparatus. With the apparatus he focuses sunlight onto a piece of scroll paper, and it shortly smokes then bursts into flames).*

P: *(impatiently)* Well, if this is not a ruse then perhaps . . . perhaps light is really a form of fire, so you have not converted anything into anything else.

D: But if I should send the light, that you now say is fire, into an urn of oil, it becomes dark . . . where has the fire now gone? Has it become the oil which you would say is the earth?

P: Indeed . . . *(pause, stammer)* well, perhaps it is as we said quintessence . . .

D: Then as I burn the papyrus *(the paper continues to smolder)*, which is a form of the earth, in the fire, and the smoke rises into the air, and the papyrus disappears, have I not converted the earth into air?

P: *(long indignant pause)*

D: Bbbbwwwaaahahahaha . . . *(Democritus bursts into a sneering and callous laughter).*

Democritus wanted real and detailed answers to scientific questions. From Democritus we got a conceptual basis of the elements. These elements, he reasoned, must have certain *complex dynamical properties* that cause them to ultimately shape and define the behavior of matter. The multitude of various properties of ordinary matter are *reduced* to the more fundamental properties of atoms. Some elements were envisioned to be little spherical balls that could freely flow (e.g., liquids), while others had hooks and could form stiff structural bonds (metals), and still others had block-like shapes that might make regular crystalline arrays (diamond or quartz). The theory had to explain all known phenomena correctly, perhaps even predict new observable phenomena, the standard to which science holds all theories.

Of course, this was an ultra-ambitious undertaking in those days. Democritus had no microscopes, or particle accelerators, to test and validate his hypothesis. But his reductionist hypothesis implied rules and organizing principles for chemistry. Democritus dubbed the basic constituents of matter "atoms" from the Greek *atmos* (*indivisible*). Out of these basic building blocks we can construct more complex objects and the forms and shapes of all that we see. The behavior of the large-scale physical world is thus emergent from the fundamental properties of atoms.

This is a wholly modern view of the physical world, as well as one of the tasks of science. While, in Democritus's theory, certain materials could change and rearrange their structure under chemical reactions (e.g., burning them, letting them rot, or dissolving them in water), the underlying atoms were immutable, unchangeable, invariant. His theory was useful and offered a prescription for further research. Here was the basic tenet of "fundamental particles," and their role as the irreducible components of all things throughout the universe, which sculpt and shape the world through their own intrinsic properties.

Alchemists over the subsequent centuries went to work. They never succeeded in turning the element lead (Pb) into the element gold (Au), or achieving any other elemental transmutation, for that matter. In countless attempts to do so they merely rearranged elements within the many exotic compounds, but they provided the service of amassing an enormous empirical "database" of recipes and processes and properties of chemicals that formed the foundation of the science of chemistry. In this sense, Democritus's theory was tested, found to be correct, but has been so significantly enlarged in detail by later science that it ultimately proved to be more of a philosophy, a prescription to actually do the hard work of science, and not to be merely contented with a dismissive shake of the wrist, invoking "air," "fire," "earth," "water," and "quintessence," panacea for lack of a deeper understanding.

What we come into contact with on a daily basis, the "everyday matter," is the first layer of the "onion of nature." It is comprised of "molecules," which are either large or small groupings of atoms. Salt ($NaCl$), water (H_2O), oxygen in the air we breathe (O_2), and methane (CH_4), the gas we use to heat our homes, etc., are all molecules, composed of combinations of the more fundamental elements or atoms. Molecules can be broken down chemically into their constituent atoms, which can then be rearranged into other molecules. Just light a match to a certain mixture of oxygen molecules and methane molecules, and these will rapidly rearrange to form water molecules and carbon dioxide molecules, releasing a lot of heat.[2] On the other hand, sodium (Na), chlorine (Cl), hydrogen (H), oxygen (O), carbon (C), and so on, are all atoms, or "elements." These are invariant, or unchanging, in chemical reactions—they are the "fundamental particles" of chemistry.

The total numbers of these atoms never change in chemical reactions—the atom of gold (Au) cannot be changed into lead (Pb) by chemical reactions. The atoms cannot be further subdivided without doing things that aren't possible in a high school chemistry lab. To smash atoms apart, into "smithereens," takes us beyond the realm of chemistry. It takes us into a deeper layer of the onion of nature, the realm of atomic and nuclear physics, eventually into the realm of quarks, leptons, and gauge bosons. These are, today, the true "fundamental particles" of nature, perhaps to be replaced by smithereens in some science of the future.

By the mid-nineteenth century, based upon the accumulated knowledge of all the known chemical processes, the elements, or atoms, were classified according to their properties by the great Russian scientist Dmitri Mendeleev. This classification scheme is called the *Periodic Table of the Elements*. The Periodic Table was a stunning summary of the thousands of years of alchemy, chemical science, and simply messing around with matter. It represented the reduction of the virtual infinity of molecules into a simple list of approximately 100 atoms found in nature (slightly more than 100 atoms is today's count; it was significantly fewer at the time of Mendeleev; many elements, such as helium, were discovered later, and many of the heaviest elements are so radioactively unstable that they must be artificially produced in particle accelerators and are not to be found on our old high school chemistry classroom wall charts). The Periodic Table represented a pattern of repetitive chemical behavior in the properties and forms of atoms as one goes to heavier and heavier atoms. By its complexity, however, it suggested that atoms may, themselves, be further reduced and may have internal structure, and that a deeper layer of *subatomic matter* must exist.[3]

THE "PHYSICS AS AN ONION" METAPHOR

Mendeleev's table was the beginning of the modern era of the science of matter. To understand this, one must appreciate that nature is, empirically, organized much like an onion. Nature has different layers of phenomena and structures as one descends to smaller and smaller distance scales. And, going downward to shorter distances, we discover, is equivalent to going to higher and higher energy scales (higher "energy per particle"; we'll define this more carefully momentarily). Although all of nature is governed by the same underlying fundamental laws of physics, the structures of complexity that we see in nature seemingly occur at different "strata" of phenomena, like an onion, and each stratum of nature is characterized by the energy needed to probe it.

What do we mean by "the energy needed to probe it"? We have to get a little bit technical here and introduce you to the common currency in talking about energies of fundamental particles and atoms: the *electron volt* or "eV."[4] Most of the science of chemistry, that is, the amount of energy involved in most chemical reactions, lies in the range from about 0.1 to 10 eV per atom. This means that when a given atom enters into a chemical bond with another atom (or an existing molecule) to form a new molecule, roughly 0.1 to 10 eV of energy is released. This is energy that comes from the forces that produce a chemical bond between two atoms, and it is typically released in the form of light, or the energy of motion, called *kinetic energy*.

The released energy is usually converted into heat (which is the aggregate random motion of atoms in a material), but it can also be seen as the light emitted from a fire or heard as the boom from a firecracker. You can usually see the released chemical energy with your eyes because a single visible particle of light, the *photon*, carries about 2 to 3 eV of energy—after all, the light entering our visual system that allows us to see is processed by various chemical reactions in our eyeballs and our brain, and so the perception of light entirely happens at the chemical energy scale.

If we can probe molecules with a source of energy of about 0.1 to 10 eV, we can often cause a chemical reaction to occur. For example, striking a match in a mixture of methane (CH_4) and oxygen (O_2) will provoke a rapid chemical reaction—a flame—yielding carbon dioxide (CO_2) and water (H_2O).[5] The match is generating photons and kinetic energy of motion of atoms of about an eV each from its own burning (usually oxygen combining with phosphorous). These energetic particles strike the methane and oxygen and nudge them into reacting, which emits more photons. Then more and more energy is released in a chemical chain reaction, and *VAVOOM*, you might have an explosion.

The physics, that is, the motion and interactions of electrons and atoms in chemical reactions—*the stratum of the chemical reactions*—is very much independent of, or decoupled from, what is going on in other stratums of nature. For example, to analyze everyday chemical phenomena, one need not be bothered by such complications as the detailed motion of the protons and neutrons that comprise the inner atomic nucleus and that exist on much smaller distance scales than the overall size of the atom. Nuclear physics is a far different energy stratum, measured in millions of electron volts, or "MeV" (see note 4), compared to the lowly 0.1 to 10 eV stratum of chemistry. Nor need we, in studying chemistry, be bothered by the slow, lugubrious astronomical motion of the earth in its orbit around the sun. In fact, it is the relative motion of atoms and the electrons within the atoms that matters for chemistry. Thermal effects, the random motion and collisions of atoms due to heat, are typically about 0.1 eV per atom at room temperature, and they increase with temperature and therefore do have effects on chemical reactions (i.e., "cooking"). But the motion of the earth in its orbit is a common, uniform drift of the assemblage of all of the earth's atoms, producing no high-energy inter-atomic collisions. Of course, if an asteroid collides with the earth, the relative motion of the asteroid's atoms hitting the earth's atoms involves very high energies, and very serious chemistry, even nuclear physics, will occur!

While the triumph of Mendeleev's Periodic Table of the Elements formed a basis for understanding all chemical reactions, we learned in the early twentieth century that the atoms themselves are not truly elementary: they are composed of even smaller, more basic objects. To understand this we must probe into the atom. And, as it goes with probes, the probe we use to analyze something should preferably be smaller than the object we wish to probe. If the probe is bigger than the object probed, it becomes a bludgeon or battle-ax, and battle-axes don't work so well for dissecting tiny things or performing dental surgery, etc.

A simple home-brew experiment that you can perform, e.g., at a child's birthday party, will illustrate this point.

A SIMPLE HOME-BREW EXPERIMENT

Get a beach ball and a straw. Have someone blindfold you. Have your assistant take some randomly chosen small items unknown to you, like a peanut, an acorn, a coin, nuts and bolts, a few other small things, and place them on a table in front of you. Now, while still blindfolded, take hold of the beach ball with both hands. Holding only the ball, try to use it gently as *a probe* of the small objects on the table, the peanut, the acorn, etc. Can you discern which little object is which, while blindfolded, and coming into contact with them only through the very large beach ball? We would guess not, unless you peeked.

Next, take one end of the long straw and, while still blindfolded, use it to trace out the forms of the same small objects. Can you now discern what these objects are and which is which? When you trace out the objects' shapes you must use a little thought and a little imagination to try to figure out what each of them is—you've become both an experimentalist and a theoretical physicist at the same time—like Enrico Fermi. With enough effort and thought you can probably figure out what little objects were placed before you on the table. Perhaps you can tell a dime from a nickel, and chunk of cauliflower from a golf ball. Go ahead—try it!

One thing is clear, if not obvious, from this experiment: *a probe that is many times larger than the object to be probed does not work very well.* Holding the beach ball, we doubt you can discern a nickel from a dime, if you can even detect either of them. On the other hand, probes that are much smaller than the object itself allow us to readily resolve the object's structure—even without seeing it with our eyes. This simple principle holds true in all of the strata of the onion of nature, including the stratum of subatomic particles. To explore the structure of the unknown "something," we must construct a probe that is smaller than the "something" we seek to study.

This seems at first blush to pose what appears to be an insurmountable barrier to studying small objects, like the innards of an atom or a particle inside an atom. How can we study a particle's inner structure if all we have are other particles of the same size? Ah-ha! Here is where two of the greatest revolutions of science come to our aid: the quantum theory and Einstein's theory of relativity.

Essentially, we learn from quantum theory that all particles in nature are also waves. This seems to be a ridiculous and nonsensical paradox, but it is the mysterious and jarring reality of quantum theory. The effects of waves vs. particles for most things don't show up until we reach atomic dimensions, but they can be seen readily for ordinary light. But to be precise, a quantum state is neither a particle nor a wave—it is both at once!

Small objects can be described by quantum mechanical waves that are associated with the probability of detecting a point-like particle at any point in space and time. That is a mouthful, and the interested reader should grab a copy of our book *Quantum Physics for Poets* (Amherst, NY: Prometheus Books, 2011). However, if you can just "ride the wave" with us for a few more paragraphs, you need only accept that a wave always has a characteristic *wavelength*. The wavelength is just the familiar distance between two crests or two troughs of a wave, like a water wave. It is the quantum wavelength that tells us how big an object is when it used as a probe.

Now here is a second relevant fact about quantum physics: as we increase the energy of any particle, its quantum wavelength becomes smaller and smaller. When the wave motion approaches the speed of light, then Einstein's theory of relativity kicks in. If you double the energy of a particle moving near the speed of light, you will halve its quantum wavelength. So, investing a lot of energy in a particle makes its quantum wavelength smaller. This, in principle, allows us to make an arbitrarily tiny probe simply by accelerating a particle to arbitrarily high energies. This is the most important principle underlying microscopes and particle accelerators. The more energy in a particle, the smaller it becomes. And, by the way, you now understand why today's particle accelerators are very large: it takes a very large accelerator to put a lot of energy into a particle to make it become a smaller probe.

The wavelength of ordinary visible light ranges from, approximately, higher-energy blue light, 0.00004 centimeters (4×10^{-5} cm, about 3 eV per photon; recall that a centimeter is about a half an inch) to lower-energy red light, 0.00007 centimeters (7×10^{-5} cm; about 2 eV per photon). A typical visible particle of a light, a photon, has a quantum wavelength in this range, with an energy of approximately 2 to 3 eV. Objects larger than about 0.0001 centimeters (10^{-4} cm) can be readily probed with visible light because they are smaller than the wavelength of the light wave. You need only make a precise optical microscope to do this, and you can see little things that your eye cannot resolve.

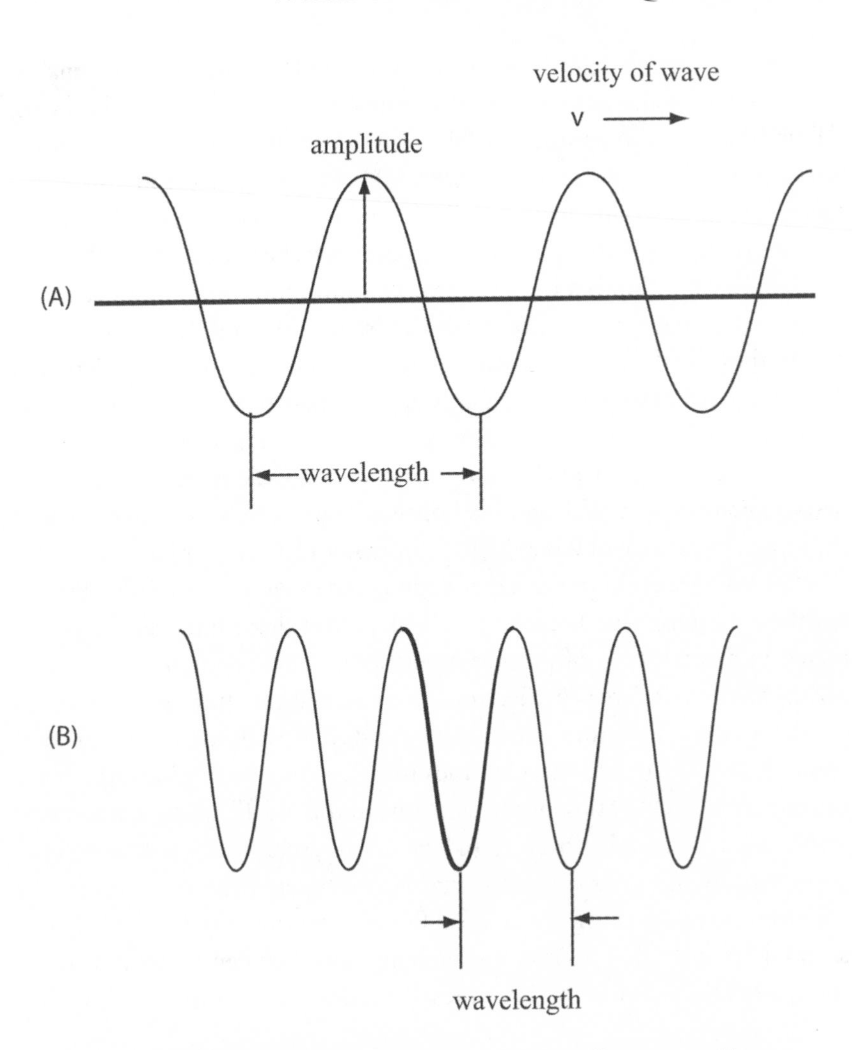

Figure 2.1. Wave and Wavelength. (A) A traveling wave. The wave moves to the right at a speed of v and has a *wavelength* (the length of one full cycle, from trough to trough, or crest to crest). An observer watching the wave travel by would see a *frequency* of v/(wavelength) crests, or troughs, passing by per second. The *amplitude* is the height of a crest above zero. (B) As we increase the energy of a quantum particle-wave, the wavelength becomes smaller.

However, visible light falters when it is used to study structures smaller than this size scale, such as the tiny components found inside the living cell of a biological organism. Visible light is unable to resolve two objects of much less than 0.00001 centimeters (10^{-5} cm) or smaller. You now know the reason: these objects are smaller than the wavelength of the visible photons of light—visible photons are as useless as beach balls to probe such small distance scales. No improvement in your microscope optics can ever improve the image. You could spend hundreds of thousands of dollars on a top-of-the-line Bausch and Lomb microscope, and still the tiniest denizens deep inside living organisms will only appear fuzzy or will not appear at all. Crank up the magnifying power of your microscope, and all you'll get is a bigger fuzzier image. You absolutely cannot see DNA in an optical (light-based) microscope. Visible light is hopeless to use as a probe of an atom at the atomic-size scale of 0.00000001 centimeter (10^{-8} cm) or less.

Fortunately, at shorter distance scales, even down to the atomic stratum and beyond, electrons become excellent probes. Electrons can be accelerated in a small type of particle accelerator called an *electron microscope*, giving them more energy. Electrons, too, have a quantum wavelength, as do all particles. Electrons can easily be endowed with kinetic energies of about 20,000 eV (that's 10,000 times more than a visible photon). This is the energy of acceleration of the electrons in the old TV picture tubes that could at one time have been found in any household but that have now gone the way of the horse and buggy. At this energy the electrons have a quantum wavelength of about 0.000000001 centimeters (10^{-9} cm), considerably smaller than that of visible light, and they can be used to make images of DNA, a virus, and even resolve individual atoms.

PEERING INSIDE THE ATOM

The first peek into the internal structure of the atom, the "atomic onion layer of nature," began about 50 years after Mendeleev with a British scientist named J. J. Thomson.[6] Thomson cleverly demonstrated in 1897 that certain "rays" that could be provoked out of atoms in an electric discharge tube (something like a fluorescent light tube) were actually particles. In particular, Thompson established that these particles lurked within all atoms, and he called the new particle the *electron*. Thomson is deservedly heralded as the "father of modern particle physics" for this discovery. He proved that electrons were extremely low-mass particles, weighing two thousand times less than the atom itself, and that they each carried a negative electric charge.

Thanks to the work of J. J. Thomson, atoms were now known to be full of these very lightweight, negatively charged electrons. But atoms themselves are normally electrically neutral, i.e., they have no electric charge; they can be "ionized" and lose an electron, hence they acquire the opposite, positive charge. Obviously, then, there would have to be an equal amount of positive electric charge inside the atom, balancing the negative electric charges of the electrons. But where this positive charge resided within the atom remained a mystery. In 1905, Thomson had proposed a theoretical model of the atom in which the positive electric charge is a medium that is uniformly dispersed throughout the entire atom, with the electrons embedded within it like raisins in a loaf of raisin bread.

Throughout this era a gruff walrus of a young man, Ernest Rutherford, played a prominent role in unraveling the inner properties of matter. Rutherford was a skilled and masterful craftsman and won the Nobel Prize for his work in elaborating the properties of radioactivity (which we'll describe later), and he had now become the director of the famous Cavendish Laboratory in Cambridge, England.[7]

Rutherford had grown up as one of a dozen children in a farm family in rough-and-tumble New Zealand, where he had learned hard work, thrift, and tinkering with technological innovation. As a child he'd played with clocks and made models of his father's waterwheels, and by the time he was a graduate student he was investigating the physics of electromagnetism. He had managed to devise a detector of wireless (radio) signals even before Marconi began his famous experiments that led to the wireless telegraph. When a scholarship brought Rutherford to the Cavendish Laboratory, he hauled his wireless device along to England and was soon sending and receiving signals over half a mile, a feat that impressed the Cambridge dons, including J. J. Thomson, the Cavendish director at the time. Later Thomson would declare, "I have never had a student with more enthusiasm or ability for original research than Mr. Rutherford."[8]

By 1909, Rutherford and his students were shooting tiny subatomic probes, called "alpha particles," at a piece of thin gold foil. They were carefully measuring the way in which the particles were slightly deflected as they scattered off of the heavy gold atoms in the target foil. The alpha particles were produced from the radioactive decay of the element radium, which naturally accelerated them to high energies—there were no man-made particle accelerators in those days. Alpha particles were now known, thanks to Rutherford, to be very heavy compared to an electron, and with a little energy they have a very tiny quantum wavelength and are therefore capable of probing deep inside the atom.

One day something utterly unexpected happened. The alpha particles were usually deflected only slightly by their passage through the gold foil, and the scientists were measuring in detail this "forward scattering." They decided, simply as a sanity check, to see if there was any "backward scattering." To their astonishment they found that one in 8,000 alpha particles bounced back toward the source. As Rutherford remembered it, "It was as if you fired a fifteen inch artillery shell at a piece of tissue paper and the shell came back and hit you."[9] What was happening here? What kind of thing inside the atom was repelling the positively charged and massive alpha particle?

No one before Rutherford had devised any way of mapping the inner shape of the atom. According to the "raisin bread" model of the atom of J. J. Thomson, the alpha particles should always have bullied their way straight through the atom—always! The atom was like a big glob of shaving cream, and the alpha particles were rifle bullets. Rifle bullets would tear straight through globs of shaving cream. Imagine seeing a rifle bullet occasionally deflected and ricocheting backward upon colliding with a blob of shaving cream. Such was the observation of Rutherford and his students.

Rutherford devoted his full energies to understanding this remarkable discovery. According to his detailed calculations there was only one way that any alpha particles could ever be deflected backward. That could only happen if the entire mass, and positive charge, of the atom was concentrated into a tiny volume in the center of the atom—the "atomic nucleus" was discovered! The nucleus's hefty mass and large positive charge could repel the positively charged alpha particles that came within its range and deflect them through a large angle, even kick them backward. It was as if within the glob of shaving cream there were dense, hard ball bearings that could cause bullets to collide and deflect. The electrons were orbiting this dense central charge of the atom. The raisin bread theory of the atom of J. J. Thomson was tossed in the trash-bin. An atom now resembled a tiny solar system, with miniature "planets" (electrons) orbiting a dense "dark star" at the center (nucleus), and it was all held together by electromagnetism.

Further experiments indicated that the nucleus was indeed tiny—one-trillionth of the volume of the atom—even though most of the mass of the atom, more than 99.98 percent of it, resided in the nucleus. At the time of this discovery, within this tiny solar system model of an atom, all the classical laws of physics were still thought to be rock solid, just as in the macroscopic solar system with the sun and its planets. The same laws of classical physics were believed to work in the atom just as they did everywhere else—until Niels Bohr showed up.

THINKING THROUGH THE ATOM

Niels Bohr[10] was a young theoretical physicist from Denmark who was studying at the Cavendish Lab, and he happened to attend a lecture by Rutherford. He was immediately captivated by this new atomic theory of electrons orbiting nuclei. Bohr arranged to visit the great man for four months in 1912, while Rutherford, at the time, was working in Manchester. Sitting down and thinking about the new data, Bohr quickly perceived something profoundly significant about Rutherford's planetary model of the atom. *It was a complete disaster*, according to the known laws of physics!

Bohr realized that, in their state of rapid circular motion about the nucleus, electrons would radiate away all of their energy in the form of electromagnetic waves very quickly. Like the swoop of a seagull into the sea, the electron orbits would quickly shrink to zero, within a tenth of a millionth of a billionth of a second. The electrons would spiral down into the nucleus. This would make the atom, ergo all of matter, chemically dead and the physical world as we know it impossible. The exact classical equations of electricity, due to Maxwell and based upon Newtonian physics, spelled disaster for the atom. Either the model had to be wrong, *or the venerable laws of classical physics had to be wrong*.

Bohr applied himself to understanding the simplest atom—the hydrogen atom—which would have a single electron in orbit around a positively charged nucleus consisting of a single positively charged particle called a proton. Thinking about the new quantum ideas that were in the air, that particles are also waves, young Bohr was led to propose a very novel idea. He argued that only certain special orbits can ever happen for electrons in atoms because the motion of electrons in these orbits must be like that of waves. These are like the natural wavelike motion of a ringing bell or a Chinese gong, with a dominant lowest tone, or *mode*, and a sequence of "overtones" or "harmonics." The lowest mode, what we mostly hear when the bell tolls, would be the one with the least amount of energy, corresponding to the wave motion where the electron is moving closest to the nucleus. In this lowest orbit the electron cannot radiate away anymore energy, because this is the state of lowest possible energy for the electron motion—the electron has no lower energy state into which to go. This special orbit is called the *ground state*. This is one of the hallmarks of quantum theory: atoms cannot just collapse into nothingness and are actually supported by the quantum wave motion, leading to the existence of a ground state, the state of lowest possible energy.

In three papers published in 1913, Bohr articulated his audacious quantum theory of the hydrogen atom. Each of the atom's magic harmonic orbits is characterized by a certain energy. An electron emits a definite amount of radiation when it "jumps" from an orbit of higher energy down to one of lower energy. It emits a photon, the particle of light, whose energy is given by the difference of the energy of the two orbits. With billions of atoms doing this at the same time, we see bright and unique colors for the emitted light, the photons all having exactly the same energies. Bohr put his theory to work and calculated the wavelengths of all the emitted photons, the colorful "spectral lines," seen in a spectroscope from hot glowing hydrogen gas. His formula worked perfectly! Electrons now moved in "Bohr orbits," or "orbitals," within the atoms.

None of this made the slightest sense in the familiar framework of Newtonian-Galilean physics. It required sweeping changes in our understanding of physics and the further development of the radical new ideas of quantum mechanics. In any case, atoms were indeed seen, by now, to be made of still smaller objects: the electrons and the atomic nucleus, and the rules of motion, the relevant laws of physics, were now totally new and totally quantum mechanical.

QUANTUM WAVES AND COSMIC RAYS

The quantum wavelike behavior of all matter was established within the first several decades of the twentieth century through numerous experiments, and the quantum theory was cobbled together (see note 3). By accelerating particles, we can shrink their quantum wavelengths. The overall characteristic "size of the atom" is determined by the "Bohr radius," which is the size of the quantum wave of the electron ground state of hydrogen and is about 0.000000005 centimeters (0.5×10^{-8} cm). The atomic nucleus is very much smaller, about one-hundred-thousandth the size of the Bohr radius. To explore these much smaller distances required much more energetic probes. High-energy particle accelerators would not arrive on the scene until the 1950s.

Nature, however, provides one extremely high-energy source of particles. These are very energetic cosmic rays that bombard Earth coming from deep outer space.[11] The cosmic rays are produced by violent processes in distant star systems, such as supernovae, pulsars, and active galactic nuclei. They are steered on the voyage to Earth by galactic magnetic fields, and their sources generally cannot be identified. The energies of cosmic rays extend way up beyond the highest energies of any particles we have ever seen or produced in the laboratory, higher than the LHC itself. The highest-energy cosmic rays ever detected have about 100,000,000,000,000,000,000 electron volts of energy (That's 10^{20} eV; for comparison the LHC design energy is 14 trillion eV, or 1.4×10^{13} eV), but these cosmic rays are extremely rare, only one of them passing through one square mile of sky every century! However, cosmic rays of energies up to about 1,000,000,000,000 eV (1,000 GeV; that's 1,000 Giga-eV, and 1 Giga = 1 billion) are sufficiently abundant that they can be put to good use to act as scientific probes—even to discover new particles.

Typically, cosmic rays, mostly protons and some heavy atomic nuclei, collide with the nuclei of nitrogen and oxygen high up in the earth's atmosphere, perhaps 10 to 20 miles up. These collisions smash the nuclei apart and send other particles as debris downward, into the atmosphere. This typically generates a plume of electrons from all of the subsequent ionization and more collisions of debris particles with other atoms. Long-term exposure to this radiation would be harmful, but we are protected by the atmosphere at ground level. Occasionally some new particle could, in principle, be produced, and it might be detected all the way down at the surface of the earth, provided it is a deeply penetrating particle. Some experiments place detectors high up on mountaintops or in balloons to try to detect less deeply penetrating particles.

From the 1930s, and even beyond the 1950s, when particle accelerators finally became available, most of the early discoveries of new elementary particles came from cosmic ray experiments. And, to this day, cosmic rays continue to serve us well in providing information that is hard to obtain from accelerators. Most recently, the masses of neutrinos were established using cosmic rays as neutrino sources in 1995 (see chapter 10).

The energy stratum of the nucleus is measured in terms of millions to hundreds of millions of electron volts. To unravel the nucleus required probes of hundreds of millions of electron volts, so people in the 1930s and 1940s turned to exploit cosmic rays.

THE MYSTERY DEEPENS: WHAT HOLDS THE ATOMIC NUCLEUS TOGETHER?

By the mid-1930s, building upon Rutherford's discovery and the new quantum theory, it was realized that the atomic nucleus is composed of particles called protons and neutrons. The proton and neutron are very similar particles, each having about the same mass, but there is a big difference: the proton has a positive electric charge, while the neutron is electrically neutral. Hydrogen has the simplest nucleus that consists of a single proton, but all heavier nuclei are made of combinations of protons and neutrons, just as molecules are made of atoms. For example, the normal helium nucleus contains two protons and two neutrons.[12]

Nuclei, like helium containing two or more protons, can only be held together by a very strong force—which is simply called the *strong force*. This nuclear-binding strong force has to be ultra-strong because the protons each have positive electric charges and therefore repel one another electrically. The nucleus of an atom like helium would instantly fly apart unless an overwhelmingly strong force compensated for this electrical repulsion and bound the protons, together with the neutrons, into the compact nucleus. Indeed, a nucleus like uranium, with 92 protons, is very unstable because of the enormous repulsive electrical forces of so many protons. Uranium therefore has many *isotopes*, such as U^{233} (where 233 = 92 protons + 141 neutrons), U^{235} (92 protons + 143 neutrons), U^{238} (92 protons + 146 neutrons), etc. Notice that we can package more neutrons into uranium because they are electrically neutral, and even help the binding together of the 92 protons. The strong force is about 10,000 times stronger than electromagnetism, and it can hold nuclei together up to about 100 protons.

Very heavy nuclei with lots of protons are generally unstable due to this electric repulsion. They undergo fission (spontaneously break apart) into lighter nuclei.

Forces, in our quantum world, are actually generated by particles. The force between two objects, like a proton and a proton, is caused by lighter particles that jump to and fro between the two protons. The repulsive electrical force is caused by the jumping of photons, the particle of light, back and forth between the protons. The strong force had to be due to something else.

A particle responsible for the strong force was predicted by the Japanese physicist Hideki Yukawa,[13] in 1935, based upon the known properties of the atomic nucleus. Yukawa reasoned that the force of electromagnetism is comparatively long range—the electric force between charged particles decreases "slowly," by the inverse square law (it falls like $1/r^2$ where r is the distance between the two particles). This inverse square law arises because the photon is a massless particle and can easily jump between nearby or distant electric charges. The force of electromagnetism is also somewhat feeble, because the "jumping probability" in quantum theory involves a small number, essentially the (square of the) electric charge (see the Appendix). This gives rise to the electric force that binds electrons to the positively charged protons in the nucleus.

On the other hand, an atomic nucleus is very small, a typical radius of about 0.0000000000001 centimeters (10^{-13} cm), about one hundred thousand times smaller than the electronic orbits that define the chemical size of the atom. This arises in part because of the much larger masses of protons and neutrons than electrons, but also by the strength of the strong force that overcomes the electric repulsion between protons. Furthermore, the nucleus is quite compact, requiring that the particle of the strong force need not produce a long-range or inverse square law force (which would have been detected outside the nucleus), but rather it is a short-range force. Yukawa realized that this required a new particle that could hop back and forth between protons and neutrons, causing the strong force, and the new particle would need to have a mass of about 100,000,000 eV (100 million eV, or 100 MeV; see note 4) to account for the short range of the new force.

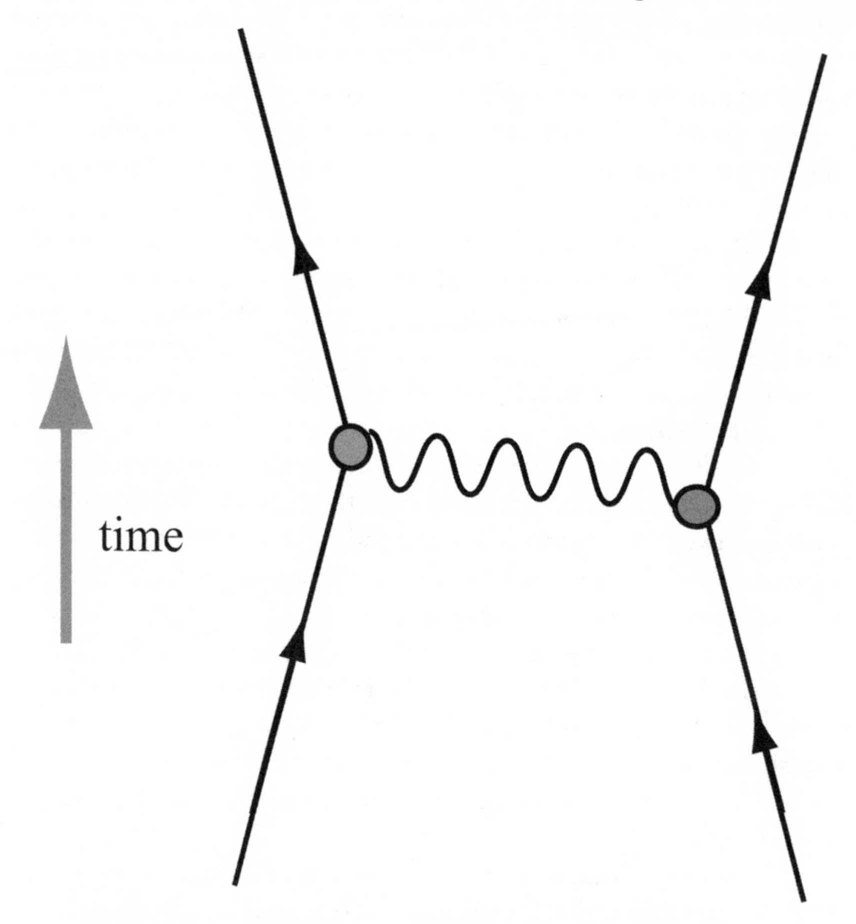

Figure 2.2. Forces Arise as the "Exchange of Particles." The force between two particles arises from the "exchange of particles." Two electrons, or any electrically charged particles, interact by exchanging photons, which are the particles of light. A proton and neutron strongly interact by exchanging pions.

This is a tall order, but it certainly pointed the particle searchers in the right direction. And remarkably, in 1936 a particle with a mass of 100 MeV, called the *muon* (pronounced mew-on), or μ, was discovered in 1937. It seemed to be the thing predicted by Yukawa, but people soon realized that the muon was a case of false identity—the cops had arrested the wrong guy. The muon was discovered by using cosmic rays that (somehow) produced it 10 miles up in the sky. The muons then traveled to the surface of the earth where they could be detected. The reason the muon was initially thought to be the particle Yukawa had predicted (called the pion [pie-on], or π) was because it had Yukawa's predicted mass. But the muon did not interact strongly enough with protons and neutrons to be a pion since it could travel all the way to the earth's surface (muons only interact electromagnetically, or through the weak force). This new particle definitely was not the agent of the strong force, as predicted by Yukawa. In fact, the muon seemed to be a mere carbon copy of the electron but 200 times heavier, with a lifetime of about 2 millionths of a second (whereby the muon "decays" into an electron and a pair of neutrinos).[14]

Almost all pure physics research was interrupted by World War II, as the world's scientists were redirected to serve military needs. The quest for Yukawa's pions could resume only after the war. In the meantime humans had conquered the atomic nucleus, with its strong force—and unleashed its fury.[15]

CHAPTER 3

WHO ORDERED THAT?

In 1947, the pions, predicted by Hideki Yukawa to explain the strong nuclear force, were finally discovered by using cosmic rays. This vindicated Yukawa's ingenious theory, for which he later won the Nobel Prize in 1949.

The pions arrived in three types, distinguished by their electric charges, π^+, π^-, and π^0, where the superscript refers to the particle's electric charge. We often refer to π^+ and π^- as the "charged pions" and π^0 as the "neutral pion."[1] The strong force, which welds the protons and neutrons together to build the atomic nucleus, arises as Yukawa had theorized through the "exchange of pions," hopping back and forth between protons and neutrons as quantum fluctuations. The picture of the atom and its nucleus was now complete.

Soon there would be particle accelerators, and numerous "elementary particles" emerged from experiments. Most of the multitude of new objects were strongly interacting, that is, they "felt" the strong force, and they interacted strongly with the pions, protons, and neutrons. It was also discovered that the proton, the neutron, the pions, and the long list of new strongly interacting particles, were not point-like objects but actually had finite sizes of about a hundredth of a trillionth of a centimeter. In the extreme cases some new particles were discovered that had lifetimes as short as the time it took light to transit their finite diameters. The nascent world of particle physics was never more confusing and chaotic than in the 1950s as the first higher-energy particle accelerators came online.

Throughout this time, the poor muon seemed to be an oddball, an almost uninvited guest at the dinner table.[2] The muon has a mass about 200 times that of the electron. It "decays" (through the weak interaction) into an electron and two very difficult to observe particles called neutrinos, living a mere two millionths of a second when at rest in the laboratory. Otherwise, the muon seemed to play no particular role in anything else, pointless in the fabric of nature. Its serendipitous and seemingly random appearance had elicited the famous quip by I. I. Rabi, "Who ordered that?"[3]

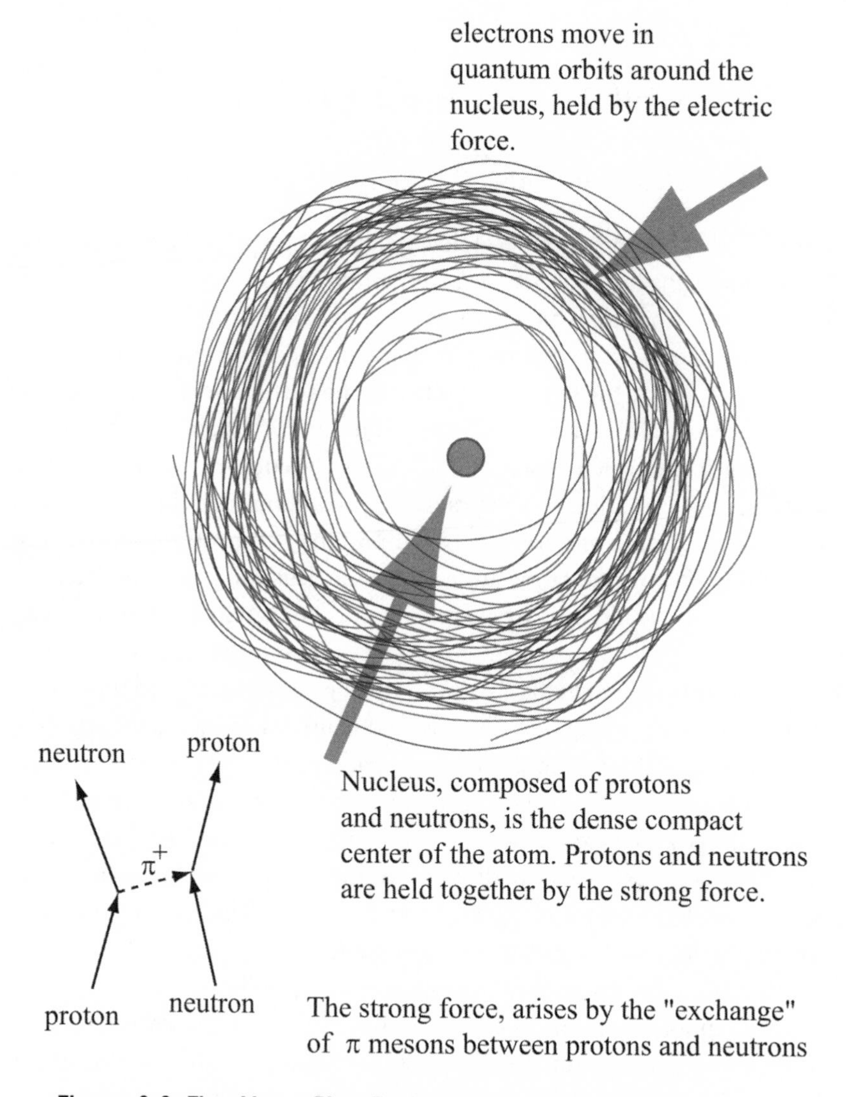

Figure 3.3. The Atom, Pion Exchange, and the Atomic Nucleus. An atom consists of the cloudlike motion of electrons about a dense nucleus containing protons and neutrons. The nucleus is held together by the exchange of pions that hop back and forth between the neutrons and protons.

We could retrace the long and winding road taken by particle physics from 1947 onward. There followed the era of the 1950s and 1960s when powerful new accelerators and various national laboratories came along. At one time there was a new energy frontier particle accelerator every few years or so, and a plethora of new "particles" and particle phenomena were discovered. Yes, we could stroll down memory lane and recount all of the history and structure of the Standard Model. Your eyes might glaze over, your eyelids becoming heavy. Rather, we'd like to veer off that traditional litany and do something a little different. We want to hop, skip, and jump to the Higgs boson as quickly as we can, to actually delve in and try to explain it to you in a way that is as close as we can get to how physicists understand it.

Indeed, by the time physicists understood the details of the forces of nature, in particular the "radioactive" transmutations between these particles that involve the "weak force," it was soon realized that some kind of "Higgs boson" was a necessity. This realization mainly came from the work of one of the architects of the Standard Model,[4] a theorist named Steven Weinberg (see chap. 1, note 14). There was no other way to make particles behave the way they do, and simultaneously to have mass, without something like a "Higgs mechanism." Remarkably, one of the key ingredients to this revelation, the ingredient that mandated the theoretical existence of a Higgs boson, was revealed by the lowly muon (in concert with the charged pion, which also decays through the weak force into a muon and a neutrino). The pion and muon decays provided the major clue about the weak forces of elementary particles that would lead directly to the Higgs boson. It was in an almost incidental way that the muon revealed the essential aspect of the weak force that ultimately legislates the Higgs boson into existence. The unexpected and uninvited guest at our table, the muon, was actually a gift—perhaps this is the answer to Rabi's question as to why the muon was "ordered."

We also think that in the not-too-distant future humans will use the muon as a powerful practical tool, much like we use everything we discover in nature. In fact, muons are already providing themselves as new diagnostic tools that scientists use to study nuclear and atomic processes. In some quarters there is a fervent albeit long-shot hope that the muon might ultimately provide the catalyst needed to unleash the ultimate energy source—

nuclear fusion—through a process known as "muon-catalyzed fusion."[5] And we believe our favorite laboratory, Fermilab (or some other laboratory in another country, if the US government doesn't get its act together), will someday build a new type of high-energy particle collider—one that will literally rank as the most sophisticated *thing* humans ever built—a machine that collides muons—the *Muon Collider*.[6] There are many reasons why this is very good ideas—so let's now pop open some champagne and celebrate . . .

THE LOWLY MUON

As we've seen, people were expecting to observe the pion, with a mass of about 100 MeV, to confirm Yukawa's theory that explained the strong force between the neutron and proton and that holds together the atomic nucleus. Instead, Carl Anderson and Seth Neddermeyer, at Caltech in 1936, found the muon in the debris of cosmic rays that bombard the surface of the earth.[7] Muons are produced in the upper atmosphere when primary, very energetic cosmic rays from outer space collide with nuclei of atoms in the thin air. But even that story has a peculiar twist.

How are muons produced in these collisions? Remarkably, we know *today* that pions are, indeed, immediately produced by the cosmic rays hitting atomic nuclei in atoms of nitrogen and oxygen in the earth's upper atmosphere. This happens because the pions are *strongly coupled* to nuclei (they are the glue that holds nuclei together, after all), and the cosmic rays are essentially protons colliding with other protons and neutrons in the nuclei of atoms in the atmosphere—at the high energies of cosmic rays, this process readily ejects pions. The pions then rapidly decay into muons (and neutrinos) within about a hundredth of a millionth of a second.

Now, think of what a bizarre situation that is—Yukawa had theoretically predicted the pion, but no one had seen the pion, and no one would see it until 1947. Anderson and Neddermeyer found the surprising new particle—the muon—coming from the cosmic rays, and it had almost the exact mass that Yukawa predicted for the pion. The muon confused the heck out of everyone, as it turned out not to be the pion. But the muons Anderson and Neddermeyer had observed, unbeknownst to them, *were the by-product of pions that are readily produced (and rapidly decay into muons) at very high altitudes* in the atmosphere![8]

Now, the very fact that a muon can make it to the surface of the earth is, in part, a miracle of Albert Einstein's theory of relativity. We know that a muon is unstable, and it almost always decays within about two millionths of second (into an electron plus two neutrinos). This is actually, approximately, its "half-life"—after about two millionths of a second, there'll be about half as many muons as you started with, then in another two millionths of a second, a quarter as many, then an eighth, and so on.[9] Muons are produced about ten to twenty miles up in the atmosphere in cosmic ray collisions. So a simple calculation shows that if a muon traveled as fast as possible, at the speed of light c, then in $t = 2$ millionths of second, it would only travel about $ct = 0.6$ kilometers, less than a half mile. So, we wouldn't expect many muons to make it to the surface of the earth before they had decayed.

But Einstein told us that, for particles approaching the speed of light, *time slows down*. The slowing of time is observed by we who are sitting at rest on the ground watching the high-speed muons. The amount of slowing down of time that we observe is the amount by which the lifetime of a muon will be lengthened due its traveling near the speed of light. This effect, called "time dilation," is easily computed: we take the energy of the muon and divide by its mass (times the speed of light squared, that is, we divide by mc^2). So, if a muon has an energy that is 20 times its own rest mass energy, mc^2, then it will have its lifetime extended by a factor of 20. With enough "lifetime-extending energy," the high-energy muons can easily reach the surface of the earth ten miles below. The arrival of muons at the earth's surface is one of the many stunning confirmations of Einstein's theory of relativity. Kooks and others who want to challenge and demolish Einstein's theory of relativity, please take note: Relativity is a prime example of a "theory" that has become "fact"!

Alas, the pion cannot travel far enough to arrive at the surface of the earth to be detected. It too is unstable, but it decays in a mere one hundredth of a millionth of a second—that's a hundred times shorter lifetime than that of a muon. We would need super-energetic pions to be produced with sufficient energy to lengthen their lifetimes by 2,000 times to get them to the surface of the earth—the effects of relativity simply aren't enough to help cosmic ray pions get down to the earth's surface.

And, there's another big difference between pions and the muons that prohibits the former from making it down to the earth's surface. For the very reason that it holds the atomic nucleus together, a pion interacts very strongly with protons and neutrons. This means that when a cosmic ray makes a pion in the thin upper atmosphere, the pion is quickly reabsorbed by protons and neutrons in further collisions with atoms of nitrogen or oxygen. This, more than anything, cuts off the number of pions in cosmic rays observed at sea level. You have to go way up into the atmosphere to detect them.

Nonetheless, in 1947 the charged pions, produced by cosmic rays, were finally found by a collaboration of scientists at the University of Bristol in England.[10] Photographic plates were placed for long periods of time at high altitudes on mountains, first at Pic du Midi de Bigorre in the Pyrenees and later at Chacaltaya in the Andes Mountains. Here the photographic plates were directly hit by the primary cosmic rays, and after development, the plates were inspected under microscopes. This revealed the tracks of electrically charged particles. Pions were first identified by unusual double tracks, where one incoming track would suddenly shift direction into another outgoing track. The scientists were actually seeing the charged pion as it decayed into the muon, plus an invisible neutrino.

Recall that the pions come in three charge species, π^+, π^-, and π^0. The neutral pion decays very quickly, in about one tenth of a millionth of a billionth of a second (that's 10^{-16} seconds), into two very energetic photons, called gamma rays, $\pi^0 \rightarrow \gamma + \gamma$. This rapid rate of decay occurs because it involves the electromagnetic interaction. But the negatively charged pions decay into muons plus antineutrinos: $\pi^- \rightarrow \mu^-$ + anti-ν^0 (and the corresponding antiparticle process involves a positively charged pion decaying into an anti-muon of positive charge and a neutrino, $\pi^+ \rightarrow \mu^+ + \nu^0$). The charged pions live about 0.00000001 (that's 10^{-8}) seconds.

The charged pion decays are examples of "weak interactions." The "weakness" of the weak interactions makes the charged pion decay more slowly than the neutral one (the weak force at low mass scales of pions and muons is much more feeble than the electromagnetic force, which is why no hints of it were observed at all until the late 1890s), and therefore the charged pion has a longer lifetime than the neutral pion.[11] As we'll see shortly, these weak decays of the pion into the muon revealed the stunning property of nature that ultimately led us to the Higgs boson.

So, you see, the scientists who worked all of this out in the 1930s and 1940s had quite a few puzzles to solve (see note 10). There is a lot of physics involved with muons and pions. Once these particles were established and their detailed properties were ferreted out of the many experiments, they became the tools to take us deeper into the fabric of nature. And the plot thickened.

THE MUON IS NATURE'S PERFECT GYROSCOPE

A remarkable feature of the muon is that it provides us with a nearly perfect elementary gyroscope. Indeed the muon has spin, and the spin of a muon, once it's produced, is very stable. The spin and its electric charge causes the muon to become a magnet. So we can place muons in magnetic fields, make them go in circular orbits, and the spin of the muon will "precess" or slowly change direction, much like a toy gyroscope in the gravity force of Earth (see the Appendix under "Spin" heading).

Welcome to the quintessential quantum property of "spin." Any rotating body has spin—a top, a CD player, a ballerina or figure skater, the earth, the washing machine basin on the rinse cycle, a star, a bicycle wheel, a black hole, a galaxy—all have spin. So, too, do most quantum particles, molecules, atoms, nuclei of atoms, the protons and neutrons in the nucleus, the particle of light (photons), or electrons or muons, the particles inside of protons and neutrons (quarks, gluons), the electron, etc., all have spin.

A notable exception is the pion, which has zero spin. But while large classical objects can have any amount of spin, and can stop spinning altogether, quantum objects have "intrinsic spin" and are always spinning with the same total intrinsic spin. An elementary particle's spin is one of its defining properties. We can never halt a muon from spinning, else it would no longer be a muon. We can never make a pion spin, hence it would not be a pion anymore.

When we say that an object spins in a particular direction, we are referring to the *axis about which it spins*. Consider a toy gyroscope. Now to define the "spin direction" of the gyroscope, we take our right hand and curl our fingers in the direction of motion of the spinning mass of the gyroscope. Our right hand's thumb then points in what we call the "spin direction." Similarly, we can define the spin direction of a muon.

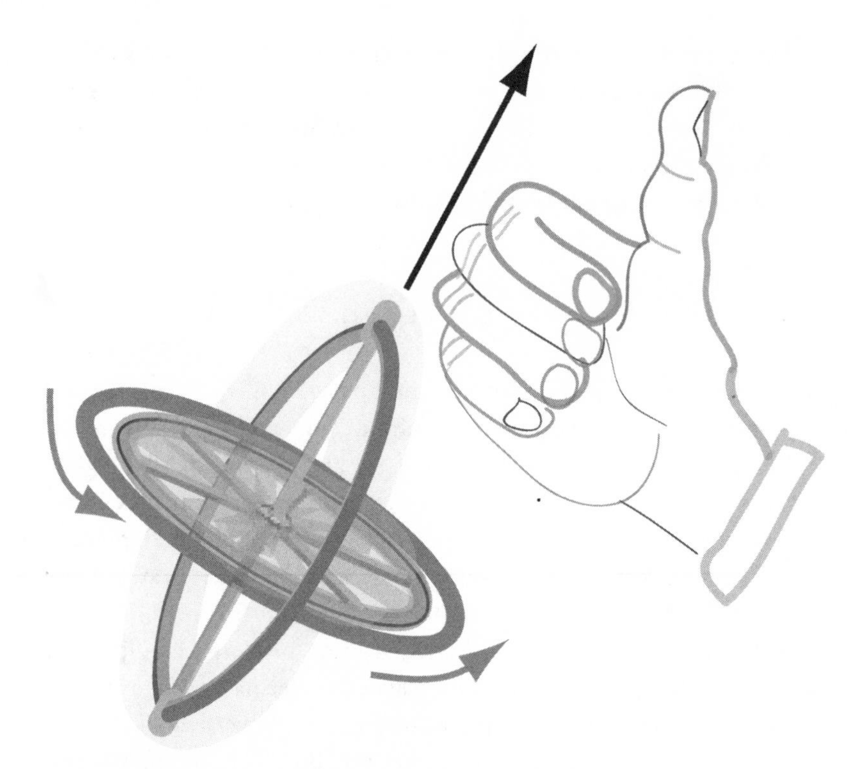

Figure 3.4. Spin of a Gyroscope. The direction of the spin (angular momentum) of a gyroscope is defined by the "right-hand rule." We curl the fingers of the right hand in the direction of the spinning limbs of the gyroscope. The spin (vector) points perpendicular to the plane of the spinning gyroscope. The assigned direction of the spin is the direction of the thumb of the right hand.

It is, however, one of the weird results of the quantum world that the spin of a muon is always either "up" or "down," once we choose any direction along which to measure the spin. So, if we choose the direction of measurement to be "east," the muon will always be observed to have exactly "east" pointing or "west" pointing spin. East-pointing spin would be spin "up" along the east direction; west-pointing spin would be spin "down" along the east direction. Quantum mechanics allows only these two possibilities, up or down, for the observed spin of an electron or a muon (or an electron, a quark, a neutrino, a proton, etc.) when measured along any axis in space.

If you don't think this is weird, then you didn't understand what we just said! A gyroscope can, at any time, have any fraction of its spin pointing east, like 29 percent, or 3 percent, or -82 percent, but an electron or muon is always 100 percent east ("up") or 100 percent west ("down")—and that's weird (this is a quantum property of spin–½ particles; see Appendix).

Spin (or, more generally, angular momentum) is a *conserved quantity* such that the total spin of an undisturbed isolated system remains forever constant. Frisbees® are a popular application of the principle of the conservation of *angular momentum*. Pilots, however, must always avoid the dreaded "flat-spin," where they can inadvertently get an airplane spinning like a Frisbee, and the conservation of *angular momentum* makes it very difficult to recover control of the airplane.

The two spin states of the resting, or slowly moving, electron or muon are easily related: we can simply rotate the muon and one spin (e.g., "up") flips into the other ("down"). It's not hard to flip the spin of an electron, in part because it is so light in mass. However, for the heavier muon these "spin flipping" effects are greatly suppressed. Once created, the muon spin will stay the same unless it encounters a strong magnetic field or experiences a really hard collision. A muon that is simply losing energy by softly scattering with electrons in a material medium will tend to preserve its spin orientation faithfully. The muon can be stopped in matter and still retains the original spin. In most experiments at low energies, the muon is a wonderful gyroscope, with very good "memory" of the spin it had when it was produced.

The existence of spin raises some fundamental questions. Recall that defining the direction of spin of a gyroscope involves using your right hand, curling the fingers in the direction of motion, and then pointing the thumb in the direction of spin. But how does nature know that it's the "right hand" and not the "left hand" that defines spin? Clearly nature doesn't care about the foolish-looking person (me) who is staring at a gyroscope, holding up his right hand as his thumb points up to the moon. As far as nature is concerned, it could just as well be his left hand. There must be a symmetry here—you should be able to consistently define spin with your left hand and the laws of physics should be the same—left should be as good as right and vice versa. That equality of properties of L and R is called "parity."

PARITY

> Now if you'll only listen, Kitty, and not talk so much, I'll tell you all my ideas about looking-glass house. First, there's the room you can see through the glass—that's just the same as our drawing room, only things go the other way. I can see all of it when I get upon a chair—all but the bit behind the fireplace. Oh! I do so wish I could see that bit!"[12]

So said Alice, before she fell through the looking glass on the mantel of her Victorian parlor to get a better view, where she had climbed to see if there was also a fire in the fireplace of the "looking-glass house." She tumbled into a new world in which the normal laws of physics were suspended—chess pieces muttered and roamed about the countryside, Humpty-Dumpty took a great fall, and "all mimsy were the borogoves, and the mome raths outgrabe."[13]

What do we see in a mirror? We see a different world, alphabetical letters reversed, the sunlight entering windows into a room that almost looks the same as ours, yet reversed, and our own image, as we are accustomed to it, but not as others see it, with that freckle and the part of the hair on the wrong side, but more or less the same. It all comes down, ultimately, to one thing—"things go the other way," as Alice said—left and right are reversed.

This left-right reversed world through the looking glass is otherwise hardly changed at all. Were we astute observers of everything that happened in this world, as methodical as Newton in trying to understand its rules and laws, what would we conclude? Would we find any difference between the laws of nature in the mirror world compared to the laws of nature in our world? Or is the "dual" world through the looking glass equivalent to ours in its most fundamental laws of physics? Would we truly find this to be a symmetry—that only the superficial things, left and right, are reversed, yet the laws of nature otherwise remain the same?

Many things are the same when viewed through a mirror. A ball with no markings looks the same in the mirror. We say that such things are "invariant under reflection." But there are also many things that are not invariant under reflection. For example, our left hand, under reflection, becomes a right hand. Right and left hands are distinct from one another.

This happens because there is a sense in which we can curl our fingers relative to the position of our thumb. This relative curling sense of the fingers and the placement of the thumb defines left and right handedness.

We thus see a basic property of reflections. Some things are the same thing under reflection, like spheres or label-free wine bottles (e.g., if our hands had no thumbs and the backside of each hand was the same as the front, then left hands would be identical to right hands and we would need only one kind of glove for each hand). Some things, like our hands, are different in the mirror world, and form a *pair* of mirror-image partners. If we reflect left, we get right, and vice versa. We refer to something that is different, or not invariant, under reflections as having "*handedness*."

It's not hard to make things with handedness. A box of screws from a hardware store will usually be "right-handed." This means that a rotation of the screwdriver by the curling of the fingers of the right hand moves the screw forward in the direction of the thumb. Seen in the mirror, the right-hand rotation becomes left-handed, but the mirror image of the screw still moves forward, so the mirror-image screw is "left-handed." A left-handed screw can just as easily be manufactured in a factory, and it is completely compatible with the laws of physics—nothing violates the laws of physics to make a left-handed screw, it just takes a special order from a manufacturer, such as "Please make us 10 dozen 8-32 left-handed screws."

At a more fundamental level, molecules generally have reflection symmetries. A molecule can be invariant under a reflection, such as H_2O, which looks the same in a mirror. Or, a molecule can become a different one, having a mirror partner, when we reflect it in the mirror. A molecule that is the mirror image of another molecule is called a "stereoisomer." A stereoisomer pair contains left (levo-) and right (dextro-) forms that swap identities by reflection in the mirror (like our left and right hands). That is, dextro-molecules are the mirror images of the levo-molecules, and vice versa. The dextro- (levo-) stereoisomers will have the exact same chemical properties when they are mixed into soup with other dextro- (levo-) stereoisomers. However, dextro- (levo-) isomers will have different chemical properties when they are mixed into soup with the mirror-image levo- (dextro-) isomers.

This leads to the old story that when you first land on a distant world, Zzyxx, and are greeted by aliens who look exactly like you and are invited to a great feast with their leader, you find that you are hungry within a few hours. The food stuff on their world is entirely based upon levo-sugars, while ours is based upon dextro-sugars—we cannot digest or gain food nutrition from these mirror-image sugars. This is an accident of evolution at which, in the distant past, some organisms randomly evolved to eat dextrose on Earth and levose on Zzyxx. This fateful event propagated forward to all subsequent generations of organisms on the respective worlds. And after a moment's thought, we realize that this is a very good thing—we will not be eaten by the aliens on Zzyxx should they turn out to be cannibals!

So, we might ask, "Are the laws of physics invariant under the discrete symmetry of reflection?" Are the laws of physics in the looking-glass house, which differs only by exchanging right for left, really the same as ours? In other words, "Is parity a symmetry of the laws of physics?"

Perfect parity symmetry, both literally and mathematically, would mean that upon viewing the world, including all of its physical processes, in a mirror, as if we were in Alice's looking-glass house, would reveal the same outcomes for experiments as in our world. In the mirror world we see physical objects move around, collide and interact, and obey the "laws of physics" that work on that side of the mirror. Are these exactly the same as ours? Is there any process in nature, let us say, in how something like a pion decays into a muon and a neutrino, or how a muon decays into an electron and neutrinos, that would be different in the mirror world than what we see in our world? It seems like such a simple idea, and such a natural one, that we might lull ourselves into believing that "yes, indeed, it must be that way! How could it not? Parity must be a symmetry of physics. What else could it be?"

The pion is a "spinless" particle, meaning that it is a "spin = 0" particle—it has zero intrinsic spin. It can be considered as a perfect little sphere, like a tiny billiard ball, which does not appear to change in any way if we rotate it. The muon (and the anti-neutrino), on the other hand, has spin that is always "up" or "down" along any axis we choose to measure the spin. When a pion decays, the initial spin is zero, therefore the sum of the final spins of the muon and anti-neutrino must also be zero. That is, if we observe the outgoing muon with spin "down" along the east direction, then the anti-neutrino will have spin "up" along the east direction. This is the conservation law of angular momentum.

An extremely important experimental point, and the reason we can do this experiment at all, is that we can slow down and stop a speeding muon, and its spin is not affected—it's a perfect gyroscope, remember? Since the slowing down and stopping of the muon doesn't change its spin direction, we therefore know the exact direction of the muon spin at the instant that it was produced from the decay of the pion. We can even measure its spin, which tells us what it was at the moment it was created in the pion decay. Furthermore, the muon itself decays (in two millionths of a second) into an electron and invisible neutrinos. The direction of its electron decay product actually reveals the muon's spin, so we have a neat way to measure it.

So, we can set up an experiment to look in detail at the decay events of the pion and the muon. For this we go into a laboratory and use pions that are produced by an accelerator. We look for events where the muon comes out with its spin aligned along the muon direction of motion (we call this "right-handed" or R). And we can also look for events in which the muon spin is counter-aligned to the muon direction of its motion (we call this "left-handed" or L). Remember, since R and L are a form of handedness, like our right or left hands, the handedness of a particle is always reversed when viewed in a mirror.

In the mid-1950s, Leon Lederman and his colleagues measured the handedness of the outgoing (negatively charged) muon, produced in (negatively charged) pion decay. Let's try to guess what the answer should be for the outcome of this experiment.

If parity is a symmetry of the laws of physics, then both R and L muons should have occurred with equal probability from pion decays (quantum mechanics gives us only the probability of something happening, and it can't tell us exactly what will happen in any given event). That is, for many decay events we should get (approximately, since there are always statistical fluctuations) 50 percent L and 50 percent R muons coming out. Any given decay of the pion would have to produce a definite handedness for the muon, either L or R, and the mirror image of any particular event would have the opposite value handedness, either R or L. So any particular decay of the pion is different than its mirror image. But parity symmetry would require that things balance out over many, many decays. If parity is a symmetry, then we shouldn't be able to infer that we live on one or the other side of the mirror from pion decays. For example, if negatively charged pions produced

60 percent L muons and 40 percent R muons, over many, many decays, then there would be something wrong with parity symmetry, because in the mirror world we'd get the reversed situation, 40 percent L and 60 percent R decays. Pion decays would therefore be different in the mirror world than in our world. If that were the case, parity would not be a symmetry of the laws of physics (you might want to reread this paragraph a few times). In any case, this is how old Democritus might have reasoned it out.

So what happened when Leon and his colleagues did the experiment? Let's hear it from the master himself.

A PERSONAL RECOLLECTION OF THE DISCOVERY OF PARITY VIOLATION

> I was driving north from a long day spent in the Department of Physics. My thoughts were occupied with the discussions that started with the traditional Chinese lunch, a Columbia University physics department Friday tradition. Friday was also the day of the weekly physics colloquium and the invited speaker was usually from some distant and, of course, lesser institution, e.g. Harvard or Yale. The Columbia lunch was designed to generate a lively discussion, usually on the topic that the visitor was going to address at his colloquium. The hidden agenda was to so saturate the hapless visitor with superb Northern Chinese cuisine that his physics defenses would be weakened and his gracious hosts could more easily destroy him.
>
> Broadway was littered with the bones of destroyed Harvard professors. The drive from Columbia University's campus in Manhattan to the Nevis Laboratory in Irvington-on-Hudson usually takes about 40 minutes. Nevis, an old DuPont estate, was willed to the University much to the chagrin of the neighboring estate owners bordering the Hudson River.
>
> The Westchester affluence begins after Yonkers with villages like Hastings-on-Hudson, Dobbs Ferry, Tarrytown. On a particular Friday afternoon, January 5, 1957, my mind was occupied with the lunchtime conversation led by Columbia's leading theoretical physicist and gourmet, Professor Tsung Dao Lee. The issue was an experiment suggested some months earlier by Lee and his Princeton colleague, Frank Yang. The experiment was designed to test a centuries-old idea in a new domain

of physics. The old idea was a belief in the symmetry between the real world and the mirror world. It had long been believed that the mirror reflections of real world processes were also real world processes. Look in the mirror. A man's jacket has its buttons on the right side of the coat. In the mirror, the buttons appear to be on the left side. But there is no law against securing the buttons on the left side.

Which side gets the buttons is a matter of convention. Similarly, turning a screwdriver clockwise (as judged by the guy turning the screwdriver) advances the screw further into the block of wood. This is a "right-handed" screw, of course. By convention, screws are right-handed. The mirror image of the process makes the screw look left-handed. But again, there is no law that prevents you from going to the manufacturer and ordering left-handed screws. They may charge you too much, but the law of mirror symmetry says it can be done and experience confirms it.

Mirror symmetry has vast implications in science. It is the same as the bilateral symmetry of the human body and of so many animals. Draw a vertical line down the center of a face and the two halves are mirror images. Molecules in which the atoms have a specific three-dimensional structure often have mirror images, which have different chemical behaviors; chemists and biologists are intensely interested in these mirror relations, but in all cases, the basic symmetry, i.e. the mirror world, obeys the same laws of physics, chemistry, and biology as the real world. Until January 6, 1957.

And until Lee and Yang had published a paper questioning whether the mirror image of radioactive processes obey mirror symmetry. What makes radioactive processes different from buttons, screws, and biological molecules, is that radioactivity is a signature of the "weak force." Certain puzzling reactions seen in the study of weak force driven decays would make sense if the weak force did not respect the symmetry. Incidentally, the validity of weak force symmetry (in the spirit of Emmy Noether's theorem) gives rise to a conservation law: the Law of Conservation of Parity. Parity is a measure of the "handedness" of a system.

Earlier in the summer of 1950, Lee and Yang had suggested a number of processes that could be studied in order to verify mirror symmetry, or to disprove the symmetry. C. S. Wu, a Columbia colleague and a skilled experimenter and expert on radioactive decays, had decided then to attempt one of the experiments which involved the radioactive decay of Cobalt 60.

The talk at our weekly Chinese lunch was that Wu had been recently seeing some interesting data, that the effect she was observing indicated a failure of mirror symmetry and that the effect could be large.

It was this information that kept circulating in my head: "A large effect?" Back in August 1956 physicists collected at the Brookhaven Laboratory, conveniently situated near some of the finest ocean beaches bordering Long Island's Atlantic shore—summers at Brookhaven were a favorite for family fun and for more serious physics discussions. Hearing Lee and Yang's proposal that perhaps mirror symmetry may fail if we concentrate on only those processes controlled by the weak forces—those that produce radioactive decays—was a revelation.

The idea sparkled for two reasons. The first was that this could explain some data that seemed totally contradictory—where 99 and 7/8ths of all reactions rigorously obeyed mirror symmetry (i.e. conserved parity) and a few reactions made no sense at all. The second deep idea was the possibility that rules and overriding laws of physics may in fact depend on the nature of the forces that are involved. As a crazy example, suppose the law of conservation of energy was valid for the electromagnetic forces, the strong forces, and the gravity force, but not valid for the weak force. The stock market value of radioactivity would go through the stratosphere! Not to worry—it isn't so.

Back to the trip north on Friday evening, following the graceful weaving of the Saw Mill River Parkway, paralleling the Hudson River. Suddenly the musings on the Wu report exploded into an idea. Suddenly, I visualized an experiment one could do with the Nevis cyclotron, an experiment so simple that one could carry it out in a few hours. (In those days, experiments, even with three or four collaborators, took months to carry out—now they take decades.)

My graduate student, Marcel Weinrich, had been working on an experiment involving muons. Muons are produced in the radioactive disintegrations of certain particles (pions) produced in the Nevis cyclotron. They behave, in all measurements, exactly like electrons—except that they are two hundred times as heavy. The issue of why nature created a heavy twin made muons a favorite object for study. Little did we dream of the treasure the muons would reveal! Marcel's set up, with simple modifications, could be used to look for a big effect. I reviewed the way muons were created in the Columbia accelerator. In this I was a sort of expert, having worked with John Tinlot on the design of external pion and muon beams some years ago when I was a brash graduate student, and the Nevis accelerator was brand new.

In my mind I visualized the entire process: the accelerator, a 4,000 ton magnet with circular pole pieces about twenty feet in diameter, sandwiches a large stainless steel evacuated box, the vacuum chamber. A stream of protons is injected via a tiny tube in the center of the magnet. The protons spiral outward as strong radio-frequency voltages kick them, adding energy on each turn. Near the end of their spiral trip, the particles have an energy of 400 MeV (1 MeV = 1 million electron volts, as though the protons had been kicked by a 400 million volt battery). Near the edge of the chamber, almost at the place where we would run out of magnet, a small rod carrying a piece of graphite waits to be bombarded by the energetic protons. Their 400 million volts is enough energy to create new particles—pions—as they collide with a carbon nucleus in the graphite target.

In my mind's eye I could see the pions spewing forward from the momentum of the proton's impact. Born between the poles of the powerful cyclotron magnet, they sweep in a gradual arc toward the outside of the accelerator and do their dance of disappearance; muons appear in their place, sharing the original motion of the pions. The rapidly vanishing magnetic field outside the pole pieces helps to sweep the muons through a channel in a ten-foot-thick concrete shield and into the experimental hall where we would be waiting.

In the experiment that Marcel had been setting up, muons would be slowed down in a three-inch-thick filter and then be brought to rest in one-inch-thick blocks of various elements. The muons would lose their energy via gentle collisions with the atoms in the material and, carrying a negative electric charge, would finally be captured by the positive nuclei. Since we did not want anything to influence the muon's direction of spin, its capture into orbits could be fatal, so we switched to positive muons. What would positively charged muons do? Probably just sit there in the block spinning quietly until their time came to decay. The material of the block would have to be chosen carefully, and carbon seemed appropriate.

Now came my key thought while heading north on a Friday in January. If all (or almost all) of the muons, born in the decay of pions, could somehow have their spins aligned in the same direction, it would mean that parity is violated in the pion-to-muon reaction and violated strongly. A big effect! Now suppose the axis of spin remained parallel to the direction of motion of the muon as it swept through its graceful arc through the channel to the outside of the machine. Suppose further that the innumerable gentle collisions with carbon atoms, which gradually slowed down the muon, did not disturb this relationship between the muon's spin and its direction of motion. If all this were indeed to happen—*mirabile dictu*! I would have a sample of muons coming to rest in a block all spinning in the same direction!

Now, dear reader, listen carefully. A spinning object can be considered right-handed (say, clockwise) when viewed from the "back end," or it can be considered left-handed when viewed from the front end. Its mirror image is the identical object turned upside down. Symmetry!

However, if this object is a muon and it disintegrates, and out of one end, there appears an electron (with lots of muons, we'd get lots of electrons), then, like the classical screw, it is uniquely right-handed. If the fingers of your right hand curl in the, say clockwise, direction of the spin, the thumb gives the preferred direction of the emitted electron. Man! That is a right-handed process. The mirror image is a left-handed object.

But if the laws of physics dictate that a positive muon is right-handed, the left-handed muon in the mirror doesn't exist—the symmetry is destroyed. The question then is: "Do electrons prefer to be emitted along the direction of your thumb or can they, with equal probability, be emitted in either direction?" In the latter case, the parity symmetry is valid. Thus, the experimental issue is extraordinarily simple—measure the direction of emission of the electrons from a collection of muons, all spinning in the same direction. If the electrons are emitted equally forward and backward (relative to the axis of spin), then parity is a symmetry and you get no promotion, no fame, and no fortune. Try again. But if there is a preference, then you have established a violation of parity, or mirror symmetry. (Okay, read it again!)

The muon's lifetime of two microseconds was convenient. Our experiment was already set up to detect the electrons that emerge from the decaying muons. We could try to see if equal numbers of electrons emerged in the two directions defined by the spin axis. Hence the mirror symmetry test. If the numbers were not equal, parity would be dead! And I would have killed it! Argggghh!

It looked as if a confluence of miracles would be needed for a successful experiment. Indeed, it was just this sequence that had discouraged us in August when Lee and Yang read their paper, which implied small effects. One small effect can be overcome with patience, but two sequential small effects—say, one percent of one percent—would make the experiment hopeless. Why two sequential small effects? Remember, nature would have to provide pions that decay into muons, mostly spinning with the same handedness (miracle number one). And the muons would have to decay into electrons with an observable asymmetry relative to the muon spin axis (miracle number two).

By the Yonkers toll booth (1957, toll five cents) I was quite excited. I felt pretty sure that if the parity violation was large, the muons would be polarized (spins all pointing in the same direction). I also knew that the magnetic properties of the muon's spin were such as to "clamp" the spin in the direction of the particle's motion under the influence of the magnetic field. I was less certain of what happens when the muon enters the energy-absorbing graphite. If I was wrong, the muon spin axis would be twisted in a wide assortment of directions. If that happened there would be no way to observe the emission of electrons relative to the spin axis.

Let's go over that again. The decay of pions generates muons that spin in the direction in which they are moving. This is part of the miracle. Now we have to stop the muons so we can observe the direction of the electrons they emit upon decay. Since we know the direction of motion just before they hit the block of carbon, if nothing screws them up, we know the spin direction when they stop and when they decay. Now all we have to do is rotate our electron detection arm about the block where the muons are at rest to check for the direction of emission and therefore, to check if parity is a symmetry.

My palms started to sweat as I reviewed what we had to do. The counters all existed. The electronics that signaled the arrival of the high-energy muon and the entrance into the graphite block of the now slowed muon were already in place and well tested. A "telescope" of four counters for detecting the electron that emerged after muon decay also existed. All we had to do was mount these on a board of some sort that we could pivot around the center of the stopping block. One or two hour's work. Wow! I decided that it would be a long night.

When I stopped at home for a quick dinner and some bantering with the kids, a telephone call came from Richard Garwin, a physicist with IBM. Garwin was doing research on atomic processes at the IBM research labs, which were then located just off the Columbia campus. Dick hung around the Physics Department a lot, but he had missed the Chinese lunch and wanted to know the latest on Wu's experiment.

"Hey, Dick, I've got a great idea on how we can test for parity violation in the simplest way you can imagine." I explained hastily and said, "Why don't you drive over to the lab and give us a hand?" Dick lived nearby in Scarsdale. By 8 p.m. we were disassembling the apparatus of one very confused and upset graduate student. Marcel saw his Ph.D. thesis experiment being taken apart! Dick was assigned the job of thinking through the problem of rotating the electron telescope so we could determine the distribution of electrons around the muon spin axis. This wasn't a trivial problem, since wrestling the telescope around could change the distance to the muons and thus alter the yield of detected electrons.

It was then that the second key idea of the experiment was invented, by Dick Garwin. Look, he said, instead of moving this heavy platform of counters around, let's leave it in place and turn the muons in a magnet. I gasped as the simplicity and elegance of the idea. Of course! A spinning charged particle is a tiny magnet and will turn like a compass needle in a magnetic field, except that the mechanical forces acting on the muon-magnet make it rotate continuously. The idea was so simple it was profound. I seized the opportunity to comfort our now blubbering grad student: "Don't worry, Marcel, this experiment will make you famous!"

It was a piece of cake to calculate the value of the magnetic field needed to turn the muons through 360 degrees in a reasonable time. What is a reasonable time for a muon? Well, the muons are decaying into electrons and neutrinos with a half-life of 2 microseconds. That is, half of the muons have given their all in 2 microseconds. If we turned the muons too slowly, say 1 degree per microsecond, most of the muons would have disappeared after being rotated through a few degrees and we wouldn't be able to compare the zero-degree and 180 degree yield—that is, the number of electrons emitted from the "front" of the muon as opposed to the "back," the whole point of our experiment. If we increased the turning rate to, say, 1,000 degrees per microsecond by applying a strong magnetic field, the distribution would whiz past the detector so fast we would have a blurred-out result. We decided that the ideal rate of turning would be about 45 degrees per microsecond.

We were able to obtain the required magnetic field by winding a few hundred turns of copper wire on a cylinder and running a current of a few amperes through the wire. We found a Lucite tube, sent Marcel to the stockroom for wire, cut the graphite stopping block down, so it could be wedged inside the cylinder, and hooked the wires to a power supply that could be controlled remotely (there was one on the shelf). In a blur of late-night activity, we had everything ready by midnight. We were in a hurry because the accelerator was always turned off at 8 a.m. on Saturday for maintenance and repairs.

By 1 a.m. the counters were recording data; accumulation registers recorded the number of electrons emitted at various directions. But remember, with Garwin's scheme, we didn't measure these angles directly. The electron telescope remained stationary while the muons or, rather, their spin axis directions, were rotated in a magnetic field. So the electron's time of arrival now corresponded to their direction. By recording the time, we were recording the direction. Of course, we had lots of problems. We badgered the accelerator operators to give us as many protons hitting the target as possible. All the counters that registered the muons coming in and stopping had to be adjusted. The control of the small magnetic field, applied to the muons, had to be checked.

All of this started working, and by 5 a.m. we had "20 standard deviations" of scientific proof, i.e., proof positive, that the directions in which electrons are emitted changes with the angle, relative to the muon's spin. Our muons were all right-handed. The mirror image, a left-handed version, does not exist in our laboratory, and hence, by extension, it does not exist in any laboratory. The Conservation Law of Parity was *not valid* for this weak force process either—the radioactive decay of the muon! We had made a profound discovery in a few days of hard work, observing parity violation in the weak decays of both pions and muons. And a few hours of data accumulation. By about 9 a.m. the word had, somehow (?), spread and we began receiving calls from physicists around the nation and, soon thereafter, from around the world. The irascible Austrian Wolfgang Pauli was soon quoted, showing his shock and disbelief: "I cannot believe that God is a weak left-hander." Yes, fame, fortune, and promotion followed in due course. And Marcel got his Ph.D.!

—Leon M. Lederman[14]

Let's summarize. The result obtained by performing the experiment of negatively charged pion decay turns out to be shocking: the handedness of the negatively charged muon produced in decay is always L, that is, we always see events as in figure 3.5 (A), and we never see events as in figure 3.5 (B)!

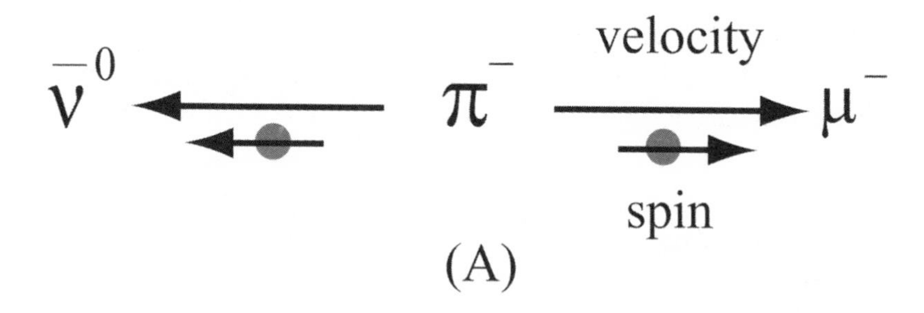

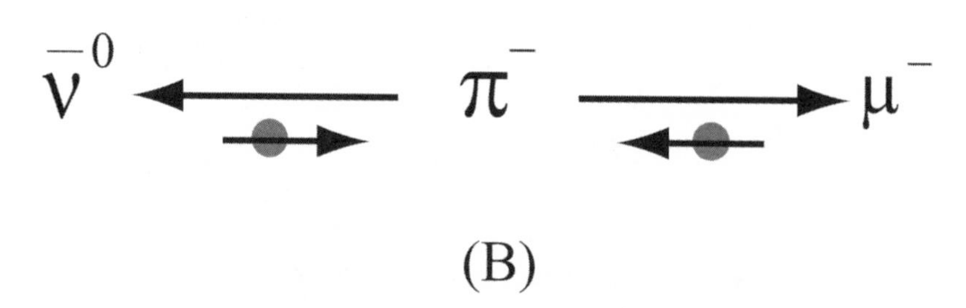

FIGURE 3.5. Parity Violation in Pion Decay. The spins of produced particles from (negatively charged) pion decays, in the weak interaction process $\pi^- \rightarrow \mu^- + \bar{\nu}^0$. In (A) the muon spin is aligned with direction of motion (right-handed muon); in (B) the muon spin is counter-aligned with direction of motion (left-handed muon). We always observe (A) in the laboratory, and we never observe (B). If we did observe (B) we could tell that we were looking at the process through a mirror.

This indeed implies that if we ever "see" a film or a DVD of a negatively charged pion decay producing an L negatively charged muon, as in figure 3.5 (B), then we can loudly proclaim: "We are seeing an image of the process reflected in a mirror! Such a process can happen only in Alice's looking-glass house. This never happens on our side of the mirror!"[15]

The mirror world with left-handed, or L, negatively charged muons coming from negative pion decay doesn't exist. (Actually, the L muon is produced instantaneously in the pion decay, but its mass flips into an R muon that balances the spin of the anti-neutrino; we'll soon have much more to say about that.) The shocking implication of the experiment is that in our world, the laws of physics contain forces and interactions that are not symmetric under parity. This happens for the class of interactions called the "weak interactions" that are producing the decay of the pion and, subsequently, the decay of the muon. Indeed, this is an example of a "broken symmetry" that occurs throughout the weak interactions, which also produce numerous other effects. The very matter out of which we are composed, hence our very existence, depends upon these feeble forces in nature, and we now learn that these forces distinguish our world from its mirror image!

Historically, until the mid 1950s, physicists had believed that parity was an exact symmetry of physics. Thus, the looking-glass world would have been indistinguishable in any movie of any process that we might ever encounter. The question of parity (P) non-conservation in the weak interactions was first raised by two young theorists, T. D. Lee and C. N. Yang, in 1956.[16] Parity symmetry was practically considered to be a bread-and-butter established fact in nature and had been used for decades in compiling data on nuclear and atomic physics. The breakthrough of Lee and Yang was the idea that the reflection symmetry—parity—could be perfectly respected in most of the interactions that physicists encountered, such as the strong force that holds the atomic nucleus together, and the electromagnetic forces together with gravity. But Lee and Yang proposed that the weak force, with its particular form of beta-decay radioactivity, might not possess this mirror symmetry.

In 1957, parity violation was discovered experimentally, by Leon Lederman, Richard Garwin, and Marcel Weinrich, by using the charged pion decay and stopped muon decay techniques we have just described. Independently, the effect was seen by Chien-Shiung Wu, using another more complex technique. It was astounding news—the weak processes are not invariant under the parity. Parity was overthrown!

Madame Wu observed the radioactive disintegration of cobalt 60 (^{60}Co) at very low temperatures in a strong magnetic field.[17] This experiment was a very challenging undertaking, requiring the heroic efforts of many groups with different expertise. The ^{60}Co is a metal out of which ordinary electrons stream, coming from beta-decay processes within the material. Wu discovered that, in the strong magnetic field, the electrons were emitted in the direction of the magnetic field (this happens because the magnetic field, at low temperatures, aligns the spins of the nuclei in the cobalt, and the decay pattern is determined by the spin of the nucleus). However, her observation was enough to conclude that there was a violation of parity symmetry. The alignment of outgoing electron velocity with the magnetic field, it turns out, is the same as a handedness, and it would be reversed in a mirror. If we saw a movie or DVD showing the electrons coming out of ^{60}Co decay counter-aligned to the magnetic field, then, again, we could announce: "This is a mirror image of the real process and does not occur in our world."

Parity is violated. Parity is not a symmetry. The mirror world of Alice through the looking glass is different in a fundamental way from ours. The lowly muon has led us to this. Perhaps that's why someone at the Chinese lunch table "ordered the muon" after all. There is a difference between left and right in our world.

And this is where the story of the Higgs boson begins.

CHAPTER 4

ALL ABOUT MASS

It's 2 a.m., the early morning hours of July 4, 2012. We have congregated in our largest seminar room, One West at Fermilab, where there is standing room only. We are here to audit two talks, one each from the gigantic experimental collaborations at CERN, known as ATLAS and CMS, talks that are being beamed in live at 9 a.m. from Geneva, Switzerland.

As the talks begin, you can hear a pin drop. The speakers nervously yet painstakingly review, in glorious detail, the data collected by the two large detectors and the complex statistical analysis required to extract a signal of any new particles from the data. They conclude at about 4:30 a.m., Fermilab time, to a standing ovation. The two CERN experimental collaborations at the LHC have just presented the official scientific discovery of the Higgs boson to the world. It is the biggest scientific discovery in the third millennium so far. We all then partied, watched the sun come up, and the next day began poring over the data in still greater detail.

For a moment the whole world stopped and gave heed to something abstract, something mind-bending, something ephemeral. From the *Hong Kong Economic Times* to the *Jerusalem Post*, from the *Fiji Sun* to the *Herald Tribune* from the *Kane County Chronicle*, to the *New York Times*, the story was carried on all the front pages of all the world's newspapers. But gradually, over the subsequent weeks, the daily news returned to the slowly recovering housing market, to a horrific mass murder, to the lingering unemployment rate, to an upcoming election, and to the usual rancid political squabbling in the US House of Representatives.

But now a much larger issue looms in the distance for human knowledge: What is the Higgs boson? Why does this thing exist? Is the Higgs a loner, or is there more to come? Are the next lessons far, far out of reach, or are we about to enter an era of rich new discoveries? What are the organizing principles? What new and mysterious reality lies beyond the Higgs boson? It all revolves around one big question.

WHAT IS MASS?

There were lots of bad jokes out there about the Higgs boson and how it gives mass to particles. For example: "A God Particle walks into a church and the priest says, 'What are you doing here?' and the Higgs boson replies, 'You can't have a mass without me.'" There was a Twitter quip by Neil deGrasse Tyson, "The Higgs discovery makes me feel heavier already. What we need instead is the anti-Higgs . . . a particle that takes mass away."[1] And, from our most reliable news source of all, the *Onion*, a more egalitarian comment: "Yeah, the Higgs boson is getting a lot of attention, but there are a lot of lower-profile bosons that are worth checking out if you get the chance."[2]

Of course, what we mean by "mass" in physics is not what is meant in other contexts, such as religion or as the term is used with "hysteria." But after some perusing of Internet sites that try to explain what physical mass is, we concluded that you'll only get confused if you start there. In part, this reflects the general confusion throughout the history of physics at arriving at a valid definition of mass. There are in fact many definitions. So let's take a fresh look at this and adopt a simple definition of mass, one that works fairly generally, and one that can efficiently launch us into the depths of elementary particles. Following Mies van der Rohe, "Less is more."

MASS IS A MEASURE OF QUANTITY OF MATTER

Yes, it's that simple. At least for the everyday objects, those that we encounter throughout our lives, mass is just a "measure of a quantity of matter." A feather or an ant has a small mass, while an automobile or an elephant has a large mass. Perhaps it is remarkable that such a simple concept is meaningful in nature, if you pause to think about it. All forms of matter, from water to steel, from putty to peanut butter and jelly, from magma to Kool-Aid® and vodka, to the interior of the sun and the cosmic rays in the depths of space—all share a common property—we can talk about how much matter they "are" in terms of something we call *mass*. We needn't specify "jelly mass" to distinguish it from "Kool-Aid mass"—it's all the same—it's just mass, which is just so-and-so much matter.

Now that's seemingly simple. But there are challenges in implementing this concept. For example, how does one measure the mass of something like a quark, which is forever trapped inside of a larger particle, together with other quarks and things called "gluons," etc., which can push and pull and actually change the mass of a quark? Just in case you've already considered some of the litany of subtleties surrounding the concept of mass in a high school physics course, then you've probably already encountered something called "inertial mass." Inertial mass has to do with the *resistance of motion* of an object *to an applied force*. This was defined by an equation written several centuries ago by Isaac Newton, "Force equals mass times acceleration." This means that, for a given applied force, an object with a larger inertial mass will accelerate more slowly than an object with a smaller inertial mass.

We would like to think the issue ends there and that "inertial mass" is all we ever have to worry about, but alas, it isn't so simple. For example, in dealing with gravitating systems, like galaxies and black holes, there are at least three types of mass, "inertial mass," "gravitational mass," and "passive gravitational mass." For quarks there's "constituent mass," "current mass," and, more generally, "off mass-shell mass." Then in discussions of relativity there's "transverse mass," "longitudinal mass," and "rest mass," etc. These definitions are all very technical and refer to specific instances in which even the best-trained physicists can become confused about energy content, motion, interactions with other stuff, and the simple idea of plain old inertial mass just doesn't cut it.

For now, let's be naive and simple-minded and forget these nasty complications and stick with plain and simple "mass is a measure of quantity of matter." The jolting revelations will come later.

MASS IS NOT WEIGHT

One reason that mass is generally a confusing concept for many people begins immediately with a trip to the moon. Most people think that the simplest measure of the quantity of matter (usually that which is their own quantity of matter, including their left foot, their head, that slight circumferential inner tube of abdominal fat, and perhaps the Big Mac® still in their stomach that was consumed at lunch, i.e., all that faces them every morning on the bathroom scale) is actually their *weight*. But, no doubt, if you took Mr. Naylor's physics class in high school, you remember some admonishment such as: "Do not confuse weight with mass! *Weight is not mass and mass is not weight.*"

In 1969, on the Apollo 11 lunar mission, humans first reached the nearest orb to their own world: the moon. When Neil and Buzz first jumped off that ladder on July 21, from their spaceship to the moon's surface, they were featherweights. Their bathroom scale weight on the moon would have been a mere 30 pounds if they'd weighed 180 pounds on Earth. They had experienced the multi-billion-dollar diet of departing from Planet Earth onboard a *Saturn V* rocket and successfully navigated to the lunar surface where the force of gravity is one-sixth of that on Earth. But—hold on—while their weight loss was a spectacular 83 percent, apart from a lean diet of mulched meat, veggies, and Tang orange liquid, their body's inertial mass was no different on the moon than on Earth. The quantity of matter that is Neil or Buzz had not changed in traversing the 200,000 miles to the moon.

The catch, of course, is that weight is a *force*, and force is not mass. While the net force of gravity that an object experiences depends upon its mass, it also depends upon the strength of the gravitational field (this is Newton's way of thinking about gravity). The gravitational field at the surface of the moon is one-sixth of that on the earth, and hence the force experienced by Neil and Buzz is one-sixth of what they experience on the earth. If NASA had shipped a bathroom scale to the moon they would have read weights that were only one-sixth of what they would read on the earth. Bathroom scales measure the *force* of gravity acting on you, and not your mass. But the quantity of matter, and hence inertial mass, that Buzz and Neil are made of is not changed by an expensive trip to the moon.

Incidentally, the extra cost of shipping the bathroom scale to the moon would have been about four million dollars (in 1969!).

MEASURING MASS CAN BE TRICKY

The best way to measure mass, in anyone's gravitational field, such as here at home on Earth, is one you have surely witnessed and that dates to the ancients. It is to use a "balance scale." A balance scale simply compares the mass of some object, let us a say a nugget of gold, to a standard predetermined quantity of matter. To establish standards, we could select our delegates and send them to a stuffy scientific conference somewhere in Eastern Europe, and there they agree upon a "standard of mass." For example, we legislate that, worldwide and henceforth, a cube of water that measures 10 centimeters on each side, at standard temperature (20° C) and pressure (1 atmosphere at sea level), has a mass of matter we call "one kilogram." We can compare any other amount of matter to our newly defined kilogram measure with our balance scale. If we place a gold nugget on one side of the balance and a 10 cm × 10 cm × 10 cm cube of water on the other side of the scale, we can immediately determine which is the greater mass: if the balance scale tips toward the gold, the nugget has more than a kilogram of mass; if the balance scale tips toward the cube of water, then the water has more mass. Voilà! We've thus made a very crude determination of the mass, or quantity of matter, that is the gold nugget.

And we can refine this in many ways. First, it's really inconvenient to use a cube full of water in our measuring apparatus as a counterweight standard of mass, because we'll surely spill it all sooner or later, and we have to correct our measurement for the mass of the container vessel of the water. So, we make counterweights: cleverly balance little pieces of lead against a carefully measured standard kilogram of water. We add or subtract shavings of lead from the lead side of the scale until it exactly balances. We then have our local blacksmith melt the lead and pour it into a mold to make a conveniently shaped cylindrical weight, much like the weight used in a grandfather clock. We then double- (and triple-) check that the resulting weight exactly balances the one-kilogram mass of water. This might take several tries, but eventually we'll have a conveniently shaped cylindrical lead weight to use on our scale instead of the cube of water that is exactly one kilogram of mass. We ask our blacksmith to make a dozen of these. We can then ask our blacksmith to forge, with extreme care, ten smaller weights that each contain exactly one-tenth the amount of lead. We can use the balance to check that each smaller weight is the same as the others, and then use the balance to check that ten such smaller weights exactly balance the one kilogram weight we started with. In this way we develop a "one-tenth kilogram" weight, also known as a "100-gram" weight. And we can go further, down to a "one gram" and beyond to a "decigram" (tenth of a gram), a "centigram" (hundredth of a gram) until it simply becomes too difficult to make a smaller weight. We now have a wide range of weights at our disposal to use with the balance.

So we can now do a refined measure of the weight of the gold nugget. Simply add or subtract lead counterweights until the lead weights balance the gold nugget. Then count how many weights are needed to achieve the exact balance. For example, we may find that we need a *one*-kilogram weight, *three* one-hundred-gram weights, *five* ten-gram weights, and *two* one-gram weights to exactly balance the nugget. Ergo, the nugget of gold weighs exactly 1.352 kilograms! And, if we ask our blacksmith and our local clockworker to make a much better scale, we might find that we also need *seven* one-tenth-gram weights and *four* one-hundredth-gram weights to get a precise balance. Hence the nugget, more precisely, weighs 1.35274 kilograms.

At this point we may find the air motion fluctuating about in the room, the atmospheric pressure of today's weather and moisture condensed on the scale, etc.; each causes some tiny amount of "noise" in the measurement of the mass of the nugget. So, our final precise measurement of the mass of the nugget of gold is 1.35274 ± 0.00003 kilograms, where the number after the "±" is the "error" in our measurement. It's important to understand that no experiment can measure anything to infinitely good precision, that is, there is never zero error in any scientific measurement. This, perhaps more than anything, distinguishes science from crackpot beliefs and superstitions about the world that are always supposed to be exactly true.

The main point here is that the balance scale makes a comparison measurement and not an "absolute measurement." The balance scale would work just as well on the moon, that is, we would find that we required exactly the same amount of lead to exactly balance the gold nugget on the moon, even though the moon's gravity is one-sixth of the earth's, and even though the nugget (and the lead weights) had a weight that was one-sixth as much as on the earth.

So, ultimately, we have made a comparison of the mass of a gold nugget to a certain amount of water that defines a kilogram that is needed to represent the same mass. A kilogram of gold and kilogram of lead and a kilogram of water all have the same mass—they are all the same quantity of matter. Note that instead of using our balance scale we might have used a "fish scale" that measures the displacement of a spring when loaded with a mass. The spring balances against the mass by exerting a *force* associated with compressing the spring, and thus the spring scale is measuring *weight*, not mass. So a spring scale would measure one-sixth the force on the moon.

MASS IS NOT ENERGY

Yet another more confusing aspect of mass, a confusion we hear all the time in the media when things like nuclear power or Higgs bosons and top quarks are discussed, is that "Einstein proved that mass and energy are the same thing." FALSE! This isn't true and it isn't what Einstein proved. We first need a brief digression on energy. (A more seasoned veteran of this issue can skip to the next section.)

Energy is real, yet it is sometimes seemingly intangible. Most physicists can readily define energy of any particular type, but devising a general definition isn't a simple task. In high school physics books, energy is often defined as "the ability to do work." Great! But this requires a precise definition of "work." The definition of energy becomes circular very quickly. So, for the moment, trust us that *energy* has a precise definition for all of its various forms, and certain sophisticated mathematical formulations of physics automate the process of determining the energy of something. Let's briefly consider an important specific form of energy known as *kinetic energy*.[3]

Kinetic energy is energy of motion and depends upon the mass and the speed of a moving object. It requires energy to make a massive object move, requiring more energy the more mass that the object has, and the more speed that we desire of the object.

As a simple example of kinetic energy, consider a familiar moving object, such as an automobile. Suppose the automobile has a mass typical of compact automobiles, of about 1,000 kilograms. We'll assume that the automobile is traveling down the highway at a speed of 60 miles per hour, which is approximately 30 meters per second. Physicists then compute that this automobile has a kinetic energy, or an energy of motion, equal to 450,000 energy units, called "joules." This number is derived by multiplying: (one half) times (the mass of the car in kilograms) times (the speed of the car, in meters per second) times (the speed of the car in meters per second). You have probably seen this formula expressed with the equation $E = (1/2)\, m\, v^2$. The energy units are named after James Prescott Joule, a nineteenth-century physicist who spent a great deal of time measuring and studying energy, especially when heat or thermodynamics was involved (Joule also invented electrical arc welding.) The statement that our automobile has a kinetic energy of 450,000 joules is a scientifically precise statement about the motion of the car and its kinetic energy.

For comparison, consider a completely different, and somewhat more bizarre, physical system—the motion of a pulse of protons in the CERN LHC, which is currently the world's highest-energy particle accelerator. One pulse in the LHC may contain about 3 trillion protons, about the number of atoms in a single living cell. The pulse is accelerated until it travels at 99.9999995 percent of the speed of light. We cannot use the simple formula for the automobile to compute the energy of the pulse of protons, because that formula comes from "classical physics" (the physics of Galileo and Newton), and its validity breaks down when things are traveling near the speed of light. Fortunately scientists know what to do in this case—they seize upon Einstein's special theory of relativity, and from this they can correctly compute the energy of the pulse of protons.[4] Therefore, even something as far removed from our everyday experience as a pulse of protons traveling near the speed of light in the LHC has a definite value for its energy. The pulse we have described, remarkably, has an energy (using Einstein's theory) of about 3,000,000 joules, about the same kinetic energy as that of a large loaded truck traveling at 60 mph down the highway!

Energy is a well-defined physical quantity that describes everything in the universe, and it always has a precise meaning in physics. Energy *is conserved* in every physical process, i.e., the total initial energy entering a process equals the total energy that comes out. If we had perfect energy conversion efficiency available to us, we could convert the energy in the LHC pulse to make a truck accelerate to 60 mph, or vice versa.

It's easy to get confused by this—if you already have a source of energy, such as a wind farm, or a coal-burning plant, or a nuclear power plant, then you can readily *convert* that energy to electricity to power your home or a factory. However, *you are always using up the energy you started with*, either from the wind or from coal or from the nuclear fuel in the reactor. You are not creating a net amount of free energy from nowhere. You could use the electricity to break water down into pure hydrogen gas and oxygen gas. You could then use the hydrogen as a fuel in your car, perhaps thinking it was a "clean" and energy-efficient fuel. However, it's just the same energy that you already produced with your windmills or coal-burning or nuclear power plants. It may not be so clean overall, depending upon where you got it. You don't get energy for free.

If energy could be produced from nowhere, or vanish into nothingness, we would say that *energy is not conserved*. However, in every experiment that has ever been conducted to measure such effects, we have always discovered that the total energy we begin with equals the total energy we end up with. *Thus, energy is always conserved in nature.* Of course, many things in our everyday lives are not conserved. The number of living organisms on Planet Earth or the total value of the stock market are two examples. Energy also takes many different forms. Energy is quite apparent in a moving object (kinetic energy) but less so in an object sitting at rest on top of a mountain (potential energy, which can be converted to kinetic energy as it falls). Energy is generally lost in physical processes that convert it into waste forms such as heat and sound. It can be lost in deforming materials, creating dents and crumples, which changes and rearranges the molecules in the material. Energy can be absorbed (or released) in the form of chemical energy, changing the physical state of matter from solid to liquid, or liquid to gas. Energy can stream out of a system carried by light and other forms of radiation. A large star that has run out of fuel, can shrink, converting its gravitational potential energy into light, until the gravitational energy is exhausted and the star becomes finally a *brown dwarf*, or even a *black hole*.

Indeed, it took physicists, chemists, and biologists a long time to understand that the principle of conservation of energy is exact and omnipotent. It governs everything. Even life-forms are governed by energy conservation—there is no special form of energy reserved for living things—all energy can be measured by the same units throughout the entire universe. If you could do all of the detailed bookkeeping and keep track of all forms of energy, you would find that energy is always conserved in any process whatsoever.

What we have learned is that, if the laws of physics were changing in time, then the principle of energy conservation, would cease to be true. If the forces of nature are different at one time than at another, then the amount of energy invested in a physical process will be different than the amount of energy invested in the same process at a later time. However, we've learned from many other diverse observations that *the laws of physics are not changing through time over time scales almost equal to the age of the universe.* Thus, the result of any particular physics experiment that we do tomorrow, or yesterday, or ten seconds ago, or ten billion years ago, or a thousand billion years in the future will produce the same results. The laws of physics, and thus all the correct equations in physics, are the same at any time in the history of the universe. This is an experimental fact. The laws of physics appear to be steadfast and eternal.

We have just glimpsed one of the most important relationships in nature: *Energy conservation is associated with the fact that the laws of physics do not change in time!* This is an example of something of a more general and profound significance known as Noether's theorem. This remarkable mathematical theorem relates conservation laws in physics to underlying and fundamental symmetry principles. It was proved by Emmy Noether, one of the greatest physicists and mathematicians in the early twentieth century.[5] The key point is that the nonchanging, or *invariance*, of the laws of physics is a *continuous symmetry of the laws of physics.* Noether's theorem says that for every continuous symmetry of the laws of nature, there is a conserved quantity.

RELATIVISTIC ENERGY

As we've noted above, Einstein's theory of relativity gave us a profound insight into the nature of energy and mass. Energy, velocity, and mass are related in Newton's classical physics. But in classical physics you can travel at any speed that you want. There's nothing sacred about the speed of light in Newton's physics—you are free to go faster if you have a really good rocket sled. Einstein overthrew the Newtonian concepts of space and time and discovered that nothing can go faster than the speed of light. Einstein found a new relationship between energy, mass, and velocity that was consistent with his new principle of relativity.

Newton would have concluded that the energy in an object that is sitting still is zero. But Einstein found that for a particle sitting still, with a mass m, there is nonzero energy, and he wrote down what may be the most famous formula ever written:

$$\mathbf{E = mc^2}$$

The implications of this formula are literally earth-shattering. We emphasize that mass and energy are two different things, but this simple formula informs us that mass can in principle be converted into energy, and vice versa. This equation is so famous that it regularly makes appearances on TV, T-shirts, license plates, cartoons, in Hollywood productions, on subway and restroom walls, in Broadway musicals, on doodles on ink blotters in the Oval Office, and throughout countless other venues. This formula literally unleashes all of the energy in the universe, for better or worse.

For example, suppose we could convert one kilogram (that weighs about 2.2 pounds on Earth) of mass into energy? Einstein's formula says that we'll get about 10,000,000,000,000,000,000 joules of energy. This is an enormous amount of energy, able to make a 10,000-kilogram (about 10 tons) spacecraft travel with a velocity of 99 percent the speed of light.

The conversion of mass into energy happens all the time in nuclear physics. Einstein's energy-mass equivalence tells us that the mass of the nucleus of U^{235}, the stuff used in nuclear reactors, is actually greater than the masses of the daughter nuclei and all the free neutrons that are produced from the disintegration of the U^{235}. The excess energy is what we get out of the nuclear power plant. Any process in which a conversion of mass to energy occurs, that is, in which the total inertial mass is not conserved, is a process that can be described only in Einstein's special theory of relativity. $E = mc^2$ is the formula that is most identified with the age of nuclear physics. However, it is a formula that holds for all things throughout the entire universe.

BUT WHAT IS MASS?

We've talked about several things mass isn't. It isn't weight and it isn't energy, though it can be converted into energy and vice versa, à la Einstein. And we've talked about how, in its simplest form, mass is just a measure of a quantity of matter. But what is mass? We haven't answered that question, and perhaps we've been lulled into the notion that the question is meaningless.

In fact, mass is a physical phenomenon. Something causes mass. It is "emergent" from something deeper and more fundamental. Now we enter a deeper realm, if not the deepest realm of all, that of elementary particle physics. This is a bit like the trip Alice took down the rabbit hole into Wonderland. In Wonderland the familiar concepts of things are changed. Many times the question we thought we were asking is the wrong question, to be replaced by another that seems to have no relationship to the first, but in the end answers the first question in a profound way while illuminating many more pieces of the grand puzzle. In particle physics we'll find an answer to the question "What causes mass?" We'll also find more questions.

Fasten your seat belt, turn to the next chapter, and prepare for a trip down the rabbit hole.

CHAPTER 5

MASS UNDER THE MICROSCOPE

The concept of "mass" as a "quantity of matter" applies to all things that we meet on an everyday scale, upward to the scales of astronomy and downward into the smallest objects in biology and chemistry.

For example, we can with some effort measure the mass of a water molecule, H_2O. We get, as expected, a very tiny number, about 0.00000000000000000000000003 kilograms (that is, 3.0×10^{-26}; you can see why we prefer to use scientific notation—it uses up less ink). A water molecule therefore has a very tiny mass. The aggregate masses of large things in nature, as Democritus had asserted, is the sum of the masses of all the many constituents. Isn't it remarkable that the concept of mass, invented by the ancients to describe large objects, like urns of olive oil and ingots of gold, applies just as well to tiny viruses, molecules, atoms, the atomic nucleus, and all the way down to the elementary particles, today in the third millennium?[1]

But we have learned new things about mass over the years. For example, as we've seen in the previous chapter, things that have mass can be converted to *energy*. This was the great insight that came out of Einstein's theory of relativity and is imbedded in his famous formula, $E = mc^2$. This formula tells us the energy for a stationary (nonmoving) object with mass m. A more complicated formula tells us how a particle's energy is determined even when it is moving, involving both its mass and its velocity (or more properly, its "momentum"; see chapter 4, note 3.) We can run the formula backward: if we know the energy and we know the velocity (momentum) we can work out the mass. So, even for a mysterious elementary particle, such as the Higgs boson at the CERN LHC, or the top quark at Fermilab's Tevatron, we can figure out its mass by measuring both its energy and its velocity in a large particle detector and by using the fancier formula.

Even in Einstein's theory of relativity, we are still deploying the same old idea: mass is a measure of quantity of matter, the ancient concept that we inherited from the Greeks (or earlier), and it is such a robust concept that it still holds true for the top quark or for a black hole. But, at the level of the truly elementary particles, we need to reexamine the concept of mass in greater detail. And, indeed, there are surprises.

THE MASSES OF ELEMENTARY PARTICLES

In the early twentieth century, with the new laws of quantum physics, the nature of Democritus's atoms was finally revealed and understood. Physics then began an exploratory descent to the shortest distances of nature. This became something of an exercise in opening a sequence of little Russian dolls, the next smaller one nested within the present one, and so on. First Russian doll: What's inside an atom? A: The atomic nucleus sits at the core of the atom and electrons orbit around it in quantum motion, but otherwise like planets in a solar system. Second Russian doll: What's inside the atomic nucleus? A: The nucleus of the atom is composed of protons and neutrons that are tightly bound together by the strong force. Third Russian doll: What causes the strong force? A: It is as described in Yukawa's brilliant theory, the exchange of π mesons between protons and neutrons (and here it starts to get complicated because there are many new particles associated with the strong force, and protons, neutrons, and π mesons are not truly elementary). Fourth Russian doll: So what's inside of protons and neutrons and π mesons? A: Particles called "quarks," and these are held together by "gluons" (see "Today: The Patterns of Quarks, Leptons, and Bosons" in the Appendix). And so on . . .

Altogether our list of the tiniest known Russian dolls is somewhat long—it includes the 6 quarks, the 8 gluons, the 6 leptons, and the 4 electroweak gauge bosons (the photon and W^+, W^-, and Z^0) and a not-yet-been-seen-but-surely-is-there *graviton*, the quantum of gravity—together with antiparticles, all of these comprise a complete list of all the known elementary particles (see figures A.35 and A.36 in the Appendix.) All of these are point-like, i.e., they have no discernible internal structure, so far as we can tell, and are what we mean by the "truly elementary particles."

You might then ask an obvious question in the spirit of Democritus and based upon the Russian doll experience: "OK, if there are so many of these 'elementary particles,' then what are quarks and leptons and all of these bosons, etc., made of?" To this we have no facts to offer—only theoretical speculation. We could quote the current hot theorist rock stars who say, "They're all made of strings." But in a few decades, with another generation of rock stars, perhaps there will be another speculative theory. It may say that "they're all made of smithereens." And maybe these ideas are right and maybe they're wrong—maybe we'll never know, no matter how many Discovery Channel documentaries about the Smithereen Theory are produced.

As we've said, today we have an "almost complete list of elementary particles." And that was the state of affairs on July 3, 2012. Things changed dramatically with the announcement on July 4, 2012, of a new object, which appears to be the Higgs boson. In fact, dozens of alternative theories about the Higgs boson and the Higgs mechanism were destroyed on July 4, as a veritable "mass extinction" of theories occurred. That's not a bad thing—it's progress (there's really no progress when science lapses into pure, almost religious, speculation about untestable things, like smithereens). So we now have the Higgs boson to add to our list. But what is the Higgs boson and why does it exist, and does adding the Higgs boson to our list make it complete? Many questions are raised by the Higgs boson. In short, with respect to the Higgs boson, "Who ordered that?"

When we are at the level of the elementary particles, we are exclusively and deeply within the mysterious realm of quantum mechanics. Here we find that the nature of the phenomenon of mass itself becomes a more enigmatic mystery and a greater puzzle. It becomes more exciting as well—the multi-millennia-old idea of mass merely as a "quantity of matter"—a concept that we've been using since antiquity—starts to break down.

NO MASS

In particle physics, for the first time anywhere in science, we meet something radically new: *there exist particles that have no mass*. A truly massless particle, but one that has nonzero energy, is unprecedented anywhere else in nature. These particles, by our ancient and traditional concept of mass, would have absolutely zero "quantity of matter." Yet massless particles exist—you can count them—they carry energy—so they do have "quantity of matter," though they have no mass. With particle physics, mass evolves into a new concept that is different than the simple old one of "quantity of matter" that has served us so well since antiquity in describing big aggregate objects. The whole concept of mass starts to become intimately related to the forces, and especially the fundamental symmetries, that govern all of the elementary constituents of all the matter in the universe. In this sense there now emerges an enormous difference between large everyday things and the tiniest denizens of nature.

So, to begin to understand what mass is at a deeper level, as a physical phenomenon of elementary particles, we must first understand what "masslessness" implies. What does it mean for a particle to exist but to have no mass? This is how Hamlet might have attacked the question.

We focus on light, to coin a bad pun. Light is composed of particles called photons. These particles have rather unusual properties compared to things like marbles or billiard balls. Photons are "point-like," that is, they have absolutely no internal size or structure, as far as we can see. Moreover, as we said, they are always moving, traveling at a well-defined speed called the speed of light, which physicists denote by the letter "c." The speed of light is very, very large, about 300,000 kilometers per second (186,000 miles per second). In fact, it takes only a little more than a second for a photon to travel from the earth to the moon. Because the speed of light is so great, it took a long time for people to measure it from experimental observations, and they had to resort to a lot of tricks and develop new techniques to do it. Today we know the value of c very precisely.

Photons were the first entities to reveal quantum physics: they also behave like particles and they also behave like waves—that is—they move in a wavelike manner, yet they also at the same time behave like particles—you can "count" them. This is a mind-numbing paradox, but it is true. This "dual" behavior of photons is called a *quantum state*, and all quantum states have it, and all things are quantum states. The wave-particle behavior is shared by all particles, electrons, quarks, muons, etc. This was the conclusion that became the bedrock of "quantum physics," and if you are a poet (or artist, or musician, or lawyer, or statesman, etc.), we have a book all about this profound yet enigmatic business for you (see *Quantum Physics for Poets* [Amherst, NY: Prometheus Books, 2011], chap. 2, note 3).

So, photons can behave like particles, despite their funny quantum waviness and their imperative to always travel at the speed of light. They do have something in common with marbles or billiard balls—each photon carries energy as it moves through the room at the speed of light. Photons can have a lot of energy, and then we call them "X-rays"; still more energy, and we call them "gamma rays," as when they are the product of such things as radioactive disintegration or supernova explosions. Gamma rays readily will go "tick . . . tick . . . tick . . . tick" as they are counted in a Geiger counter. It's dangerous for living organisms to be exposed to too many X-rays or gamma rays because they tend to destroy biological tissue, such as DNA. But photons can have less energy, becoming the light we see, and with still less energy, will fade off into the far red scale of visibility, becoming warm, gentle infrared light emanating from a soothing fire in the fireplace on a cold winter night, finally becoming at the lowest energy scales microwaves and radio waves.

The intriguing and novel thing about photons is that they each *have absolutely no mass*. As we have said, photons are *massless* particles. In fact, as far as what we have directly observed in the lab, photons are the only truly massless, freely moving particles (we expect there exist other massless particles, such as the hitherto unobserved particles of gravity, called "gravitons," and the gluons that bind quarks but are trapped forever inside of hadrons, so we never get to see them freely moving as massless particles through space—see figure A.37 and surrounding text in the Appendix).

"Hold on!" exclaims Katherine, who looks up from studying her law school notes in preparation for the bar exam, "Didn't that grandfatherly old man with long, bushy white hair and a pipe once say that energy is equivalent to mass? So, if a photon has energy, how can it have no mass? If it has no mass, doesn't $E = mc^2$ tell us it has zero energy? How can a photon therefore exist at all if it always has zero energy?"

Yes, Katherine, indeed the photon has no mass, but it does have energy. The photon defeats this apparent conundrum by never standing still—a photon always moves at the speed of light, and you cannot arrest a photon (in a real sense) and bring it into a stationary state of zero motion. The photon's very existence is a legal loophole in Einstein's relativity that permits massless particles to also have energy (and momentum), provided they always travel at the speed of light (see chapter 4, note 4). In fact, this goes to the core of Einstein's theory of relativity: no matter how fast we chase after a photon, it always moves away from us at exactly the same speed of light. You can never slow a photon down and place it on a balance scale to measure its mass, because its mass is zero and it always moves at c. "I see," she replies, "photons have no mass, so they always travel at the speed of light. How clever is the fine print on the legal contract of nature! But why?"

Photons are a sort-of exceptional case. Most particles have mass (in fact, all other known elementary particles at this time are massive particles, with the exception of the eternally trapped gluons and unseen gravitons) and thus, at least in principle, any elementary particle can be brought to a state of rest and will then have Einstein's famous amount of "rest energy," $E = mc^2$. But not the photon. The photon is a special particle that *can never be brought to rest*. Now, think about that for a moment. Isn't this interesting? Even if we are talking about a massive particle *at rest*, the formula for its total energy, *at rest*, is $E = mc^2$, yet this involves c, the speed of light. The speed of light is intrinsically wrapped up in all of this phenomenon of mass. It is fundamental. We call c a *fundamental constant of nature*. It governs all properties of motion, whether we are at rest or moving near the speed of light.

As we've seen when a massive object moves, relative to us, it acquires additional energy of motion, known in physicists' jargon as *kinetic energy*. It acquires just a little kinetic energy if it moves slowly. But the kinetic energy becomes greater and greater as the particle moves faster and faster. And, as the massive particle approaches the speed of light, its total energy becomes infinite. So, in fact, no massive particle can ever travel at the speed of light, because it would require an infinite amount of energy to make it do so. The photon does it by being massless, but the photon can therefore never be at rest, and all photons travel at the speed of light.

SYMMETRY IN MASSLESSNESS

The existence of massless particles raises the interesting question: Why must mass exist at all? What kind of world would we have if all other elementary particles were massless?

It certainly wouldn't be a very hospitable place in which to live. One of Einstein's results of special relativity is that time ceases to exist for objects that travel at the speed of light. That is, if a lowly photon carried a wristwatch and departed Earth from a flashlight heading for a distant galaxy, he would notice that he arrived at his destination instantaneously. The vast distances of intergalactic space present no problem for photons to traverse—from their perspective they do all trips instantaneously. No time would elapse at all on the photon's wristwatch in hopping from Andromeda to the Milky Way, or from Earth to UDFj-39546284, one of the most distant galaxies ever recorded.[2] But, alas, for our little photon there would be no time to read a good book on the trip or to catch up with some zzz's. So, too, if we were all massless particles, we would always take zero time to do everything and anything, and we would always be on the go—at the speed of light. Our world would be completely devoid of experience as we know it. What kind of life would that be? Well, we wouldn't age. But unfortunately, a world without time is non-experiential. We wouldn't age, but we also wouldn't live.

But from a purely mathematical point of view there is something very special about a world in which all particles are massless. This is a world of supreme symmetry. For example, there would be nothing to distinguish the muon from the electron in such a world—both would be exactly massless charged particles, and we wouldn't notice if we swapped all the electrons in a box with muons. When two systems have identical properties, we say they are *symmetric* to one another.

Symmetry is now known to be at the heart of our understanding of nature. In essence, we live in a world that is fundamentally governed by symmetry (and we have another book for you on that topic: *Symmetry and the Beautiful Universe* [Amherst, NY: Prometheus Books, 2007], chap. 4, note 4). But often the symmetries are hidden or appear to us as nonexistent or "broken" symmetries. This is the major lesson of the Standard Model that unifies all the forces of nature into a common logical framework. *The Standard Model achieves its unification by first imagining this supremely symmetric and un-marred world in which all particles exist without mass.* That's where we then see the unifying principles at work.

On the other hand, the real world of planets, stars, iPhones®, and humans is a world of broken-down symmetries, as though we live among the ruins of some ancient civilization. Here we see an old pillar lying on the ground and over there the keystone of an arch half buried in the mud next to a decapitated statue of Emperor Vespasian lying on its side. This is the world we encounter every day. It is a world in which things have mass and are different, as electrons are different in mass from muons. The masses of particles and atoms, and the lugubrious mass of a Jupiter, are all the symptoms of the broken symmetry of the Standard Model. The grand symmetries of the perfectly massless world of the Standard Model are hidden, just like the pinnacle of the great civilizations of ancient India, China, Central and South America, Persia, Greece, or Rome are hidden in history.

The analogy is striking when we realize that in the very earliest instants of the big bang these symmetries were fully in place and at work sculpting the future universe. In the first instants of creation all particles *were* massless, and the great vaulted towers of the symmetries of the Standard Model once stood aloft and uncorrupted. The universe expanded and cooled, and the symmetries fell into heaps of rubble, particles acquired mass, and the physics of our low-energy world of human perception emerged where the underlying Standard Model is hidden and hard to see. If by a licentious metaphor this symmetrical world was the Valhalla of Odin, then it was Götterdämmerung that broke down the symmetries and smashed Valhalla into ruins. And just as there was an agent of that event, Odin's daughter, the Valkyrie Brunhilde, so, too, is there an agent of the destruction of the symmetry of the Standard Model in the very early universe: the Higgs boson.

THE QUANTUM REALM

Another Götterdämmerung happened at the beginning of the twentieth century—the world of classical physics collapsed. Classical physics had evolved from the mists of history, to the new rational minds of Kepler, Galileo, and Newton, to Maxwell and Gibbs, and through to the end of the nineteenth century. Classical physics always involves descriptions of things that are *macroscopic*, involving collections of huge numbers of atoms. Some million, trillion atoms are contained in a single grain of sand. However, at the beginning of the twentieth century, the established and grandiose science of classical physics, with its precise predictions for the behavior of all things that are huge assemblages of atoms, like a four-hundred-year-old European monarchy, crashed down to the floor.

Through the newly refined and sophisticated experiments at the turn of the century, a revolution occurred, and the properties of individual atoms and those of the smaller particles the atoms contain came into view. The behavior of the individual atom itself turned out to be nothing like what Galileo and Newton had conceived. It was shocking and inexplicable to the scientists of the early twentieth century, who had been trained in the Galilean–Newtonian tradition of classical physics. A chaotic confusion emerged in a vast assortment of the data on atoms, but this gradually gave way to the desperate and intense efforts of the scientists to restore order and logic to this newly discovered realm. By the end of the 1920s, the basic logical framework of the new properties of the atom, which define all of chemistry and everyday matter, had been constructed.

And the logic seemed incomprehensibly illogical—but it worked and it survived many an onslaught by the doubters, including no less than the founding father of modern physics, Albert Einstein himself. Humans had begun to comprehend the bizarre new world of the smallest things, from atoms on down, that we now call *the quantum world.* The weird new quantum laws that now ruled the atom were primary and fundamental, and these new rules actually apply to everything, everywhere in the universe. We are all made of atoms and we cannot escape the implications of the surreal reality of the atomic domain, that nothing is solid, that atoms are mostly empty space, that "uncertainty" is now decreed and installed into the laws of nature.

Within this new quantum world, the concept of mass is also radically changed once we get down to the smallest denizens of nature, the "elementary particles." A new burning aspect of mass rears its head, and the notion that it's only about the "quantity of matter" has to be written into our psyche. All of this new insight into elementary particles begins in the period of the development of the new particle accelerators, beginning in the postwar 1950s. These were the world's most powerful microscopes, and they began to reveal a new layer of matter. Particles that have lifetimes no longer than the time it takes for light to transit their tiny diameters, that are smaller than the atomic nucleus, glinted and sparkled into view in the detectors of the experimentalists.

THE EMERGENCE OF QUANTUM IDEAS OF MASS

The first new ideas about mass came from the minds of theorists and came initially from outside of particle physics. This derived from the new understanding of the world of ordinary materials through the lens of quantum theory, in particular, the phenomenon by which materials that are poor conductors of electricity, like lead or nickel, become *perfect* electrical conductors—"superconductors"—at ultra-low temperatures. Yes, we said "perfect"—superconductors have absolutely zero resistance to the flow of an electrical current! This is an astonishing and ghostly quantum behavior of aggregate matter.

Superconductivity was first observed in the laboratory in the early 1900s, and the first hints of a theory were given by Fritz London in the 1930s. But it was in the mid-1950s that superconductivity was finally explained in detail by a beautiful theory of John Bardeen, Leon Cooper, and Robert Schrieffer. This, and other work by Vitaly Ginzburg and Lev Landau laid a foundation for the new quantum ideas about mass.[3]

Inside of a superconductor, like a small bar of ultra-cold lead, which can be easily constructed in a laboratory with good cryogenic equipment, the massless photon, the particle of light, becomes heavy—it acquires mass. We can actually, in principle, make photons stand still in a superconductor! It is as though one has created a mini-universe in which the vacuum state has been modified (it is filled with lead, or nickel or niobium, and is cooled to less than 2° above absolute zero temperature, that is, 2° Kelvin) and the quantum dynamics of this material causes a photon, the otherwise massless particle of light, to become a heavy particle. A superconductor allows us to become the architects of a little artificial universe in the lab, and it offers a switch that allows us to turn on or off a mass for an otherwise massless particle.

The mechanism of a superconductor can be described at different levels of detail, but it provides an insight into how mass, the quantity of matter of a particle, could be created by nature itself through quantum effects. The symmetries that are associated with the massless photon become hidden in a superconductor. The photon blends with the particles in the superconducting state to become something else, a new kind of photon with mass. This tells us that the properties of nature's *quantum vacuum* itself are inextricably wound up with the properties of particles and their masses.

Superconductivity is so well understood today that it has become an industrial tool. The enormous magnets of the Large Hadron Collider at CERN, and formerly those at the Fermilab Tevatron, use superconductivity to produce otherworldly strong magnetic fields at minimal cost in electricity. And, as a spin-off of the Fermilab Tevatron magnets, the powerful magnets used in MRI machines were born. Someday you may have a superconducting coffeemaker in your kitchen.

The underlying theoretical ideas of superconductivity were imported into particle physics by Jeffrey Goldstone, Giovanni Jona-Lasinio, Yoichiro Nambu, and others in the late 1950s to early 1960s.[4] The masses of elementary particles, at least the strongly interacting ones, the proton and the neutron, were beginning to look like a dynamical phenomenon, something that had to do with the vacuum of space itself.

IT'S ALL IN THE VACUUM

As weird as the quantum world can get, perhaps one of the strangest notions is that the vacuum itself is not empty, but rather, it is a complicated structure. The vacuum is a quantum state. We are pretty sure (not absolutely sure) that it is the state of lowest energy, called the "ground state." And all quantum states, including the ground state, can have complex features—they are not empty. For example, the ground state of a hydrogen atom has an electron orbiting the proton in a spherical cloud-like wave—it is not empty.

So, too, we've just seen that the ground state of a superconductor contains a soup of particles that effectively give the photon a mass. Our vacuum's particular structure fundamentally and inextricably influences the properties of particles. Particles are now viewed as "excitations" of the vacuum—the concepts of the vacuum of space and time together with the elementary particles become welded into one. It is as if Shakespeare's Hamlet has as much to do with the other characters onstage as with the stage upon which they perform. (Shakespeare may have been the original quantum theorist.)

HOW CAN I ESCAPE THE VACUUM?

So, there's now a new complication we must dissect—the inseparable vacuum and its relationship to matter. These are not disjointed but are united—just as the brain is united with the body. But physicists have to dissect nature, and to do it they need a tool to turn one thing off while another thing is on.

In the quest of understanding mass, the place to start is to contemplate a particle that doesn't feel the effects of the vacuum structure. This is a particle, like the photon, that can have energy but has no mass at all—a particle that always travels at the speed of light—arriving instantaneously at any destination—and thus has no experience of the vacuum along the way.

As we noted, the world in which particles are all massless would be a world of profound and elegant simplicity, and simplicity in physics comes from *symmetry*, while the world in which we, the massive particles, actually live is one of broken-down symmetry. However, as ordinary particles approach the speed of light, they, too, begin to behave much like massless particles. If you hopped on a rocket ship that could travel at nearly the speed of light, you could take the trip from Earth to Andromeda, and the time elapsed on your wristwatch could become as short as you wish, depending upon how close to c, the speed of light, you can get. As you approach the speed of light you become much like a photon, experiencing no lapsing of time as you traverse the entire universe. You become an *approximately* massless particle yourself, as seen by a stationary observer at rest in the lab! And, by examining the behavior of *approximately* massless particles in the lab, any heavy particles traveling near the speed of light, we can glimpse the world where the symmetry is restored—the world of masslessness. The effects of the vacuum become decoupled from these near-to-the-speed-of-light particles. We can, therefore, by studying approximately massless particles, i.e., very energetic particles, at least in our mind's eye, restore the vaults and towers and walls of the ancient world of symmetry, much like archaeologists reconstruct a view of an ancient city.

Let's begin a mental journey into such a perfect world—a world in which there is no mass, a divine world of perfect symmetry, where particles travel always at the speed of light. Our journey toward understanding mass is about to become quite intriguing.

PARTICLE UTOPIA

It's essential we now draw some pictures. We begin by drawing a picture of the motion of a massless particle in space and time. That's what physicists did when they started asking these sorts of questions about mass. We're going to draw good old-fashioned pictures on a two-dimensional page of paper. This will be like a map, but it must somehow display the three directions in space and also the one direction in time.

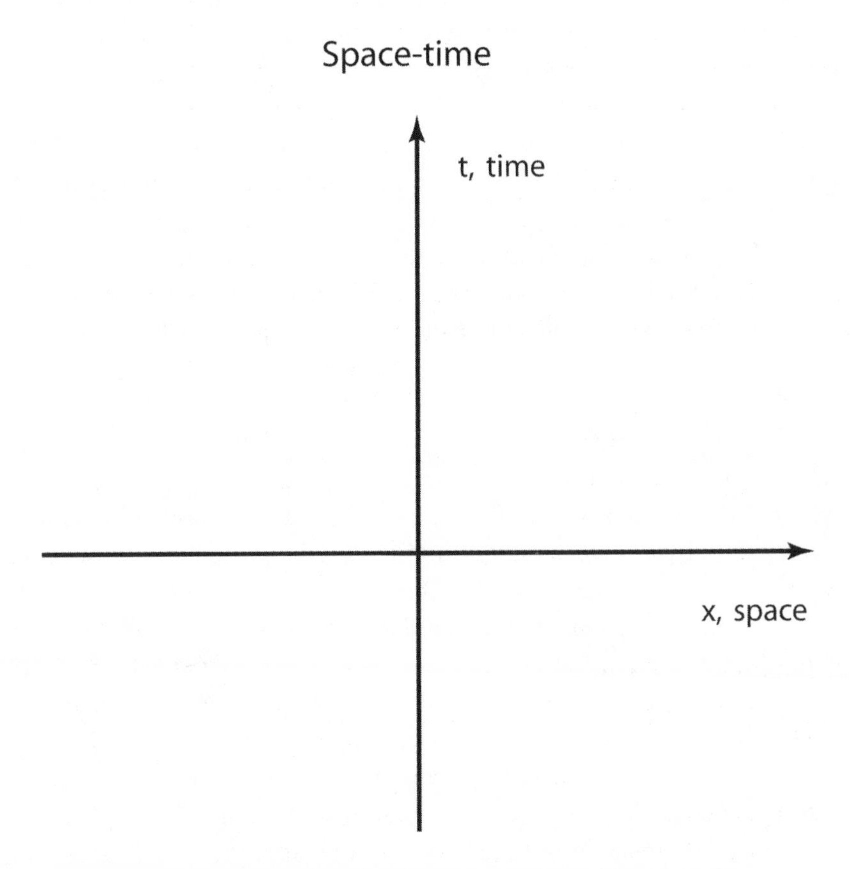

Figure 5.6, Space-Time. The horizontal axis represents all three directions in space, while the vertical axis represents the flow of time.

Alas, space is three-dimensional, and we always have a problem in rendering space—it's impossible to exactly draw three dimensions of space on a two-dimensional piece of paper. Furthermore, we will also have to use one of our two dimensions on the piece of paper to draw the fourth dimension—that of time. So, to depict space and time, we can only draw one horizontal axis to represent all of space, and let's say that this represents the east-west direction—like any map, east goes to the right and west to the left. You have to use your imagination to see the two other axes of space, one representing north-south, and the other up-down, coming out of the page and into your living room. The vertical axis in our picture represents time. In figure 5.6 we have drawn the basic map. This is a picture of a new world that we call "space-time."

Now, in nature there's something weird about time that distinguishes it from space. In space, we can always decide where we want to be. If we want to be in Antigua, Guatemala, we can hop a plane and enjoy beautiful coastlines and wonderful coffee, the colorful dress and the kite flying of the local Mayan descendants' culture. But we have no control over where we are in time. We just "are" someplace in time. Of course, when we "are" at some time, we also "are" someplace in space. In our plot, a definite time and a definite location in space is a geometrical point, a "space-time" point. A point in space-time is called *an event.*

For example, on the afternoon of July 4, 1927, there was an event at which little Billy Johnson lit off a firecracker in front of his father's hardware store on Main Street in Bedford Falls. The firecracker exploded with a loud bang. That particular instant in time at that particular point in space defined an event in space-time: the "firecracker event."

Our world is a fabric of a countless infinity of events. A mosquito stings Mrs. Fenster on her leg at her niece's ballet recital in the gymnasium of her local high school on September 20, 2003, at 2:31 p.m.; an atomic nucleus of Uranium spontaneously decays at exactly 5:23 a.m., GMT, deep in the exact center of the earth; a supernova explodes eight billion years ago in the constellation Taurus, at a distance of eight billion light-years from the earth; the light from said supernova will be detected by telescope in Chile at exactly 12:09 a.m. local time on January 12, 2015. These are just a few of the countless infinity of events that define our world. Some are in the deep past, others in the distant future. And some are related to one another, like the observation by very smart aliens many light-years away of the light emitted from the firecracker explosion by Billy Johnson. We might say that physics is the collection of events in space-time and the rules by which they relate to one another.

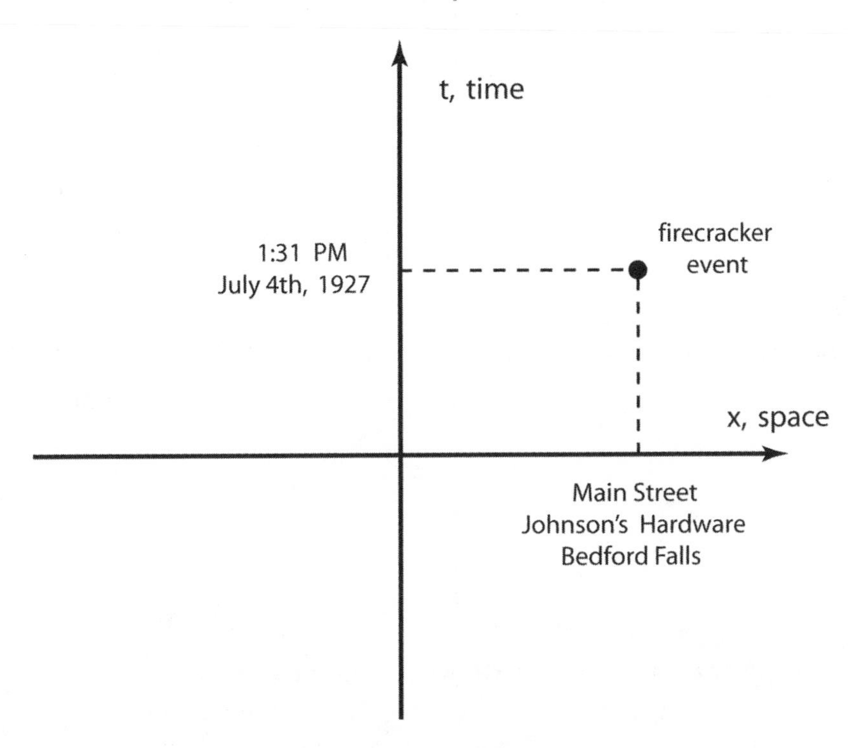

Figure 5.7. An Event in Space-Time. This event, consisting of the explosion of a firecracker, is located at some particular value of time (which we have labeled as 1:31 PM, July 4th, 1927) and at some particular location in space (which is directly in front of Mr. Johnson's Hardware store on Main Street in Bedford Falls).

Let's return to Billy Johnson's firecracker experiment. It's convenient to reset our clocks and call the time of the fire- cracker event "zero," so our "time coordinate" for the firecracker explosion is defined to be t = 0. And, likewise, we reset our space coordinates so that the location of the exact spot at which the firecracker explodes is also "zero," or, for our plot x = 0. Our space-time plot for the world then looks like figure 5.8:

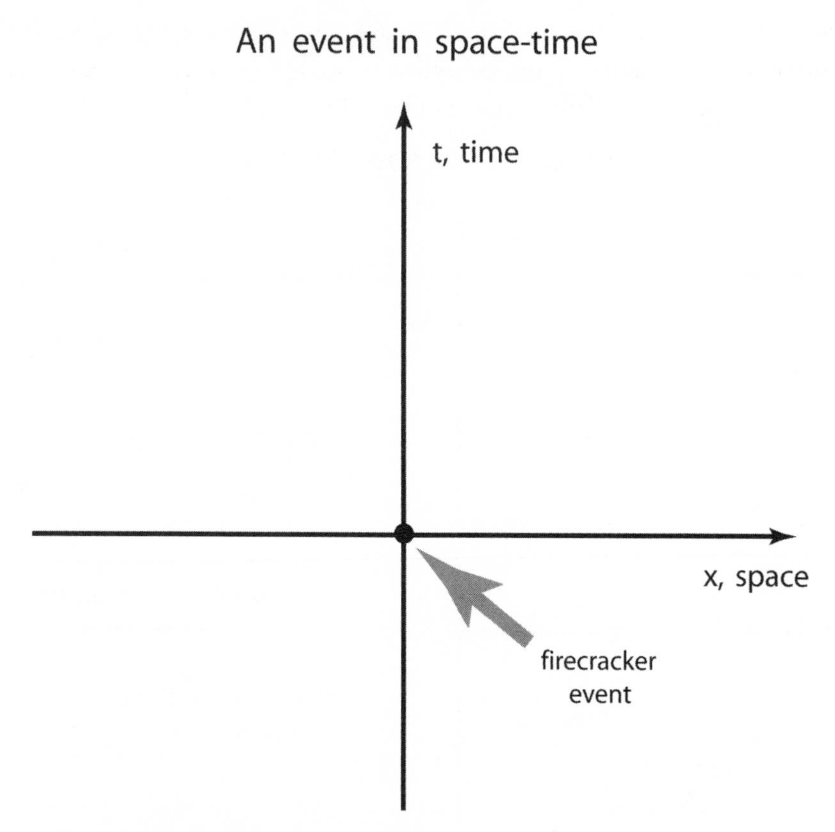

Figure 5.8. Firecracker at Origin. For convenience, we relocate the orign of time and space to coincide with the event of interest. The event, consisting of the explosion of a firecracker, is relocated to a value of time $t = 0$ by simply resetting our clocks, and a location in space, now denoted $x = 0$, by resetting our map coordinates.

Now, when a firecracker explodes, there are physical consequences. If we plot the sequence of events, they look like figure 5.9. Of course, there is an instantaneous heating and compression of the atmosphere at the event of the explosion, and this produces sound waves, actually more of a shock wave in the air, the "bang" that emanates outward triggering an infinity of other events. For example, a brief instant later the shock wave reaches Billy's brother Tommy's ears. Tommy Johnson happened to be standing a mere ten feet from where the firecracker exploded out in the street in front of the hardware store. We label this on our figure 5.9 as event (A). A brief instant later the "bang" shock wave has spread farther and reaches the ears of Mr. Johnson, who is in the backyard of the hardware store unloading boxes, an event we label (B).

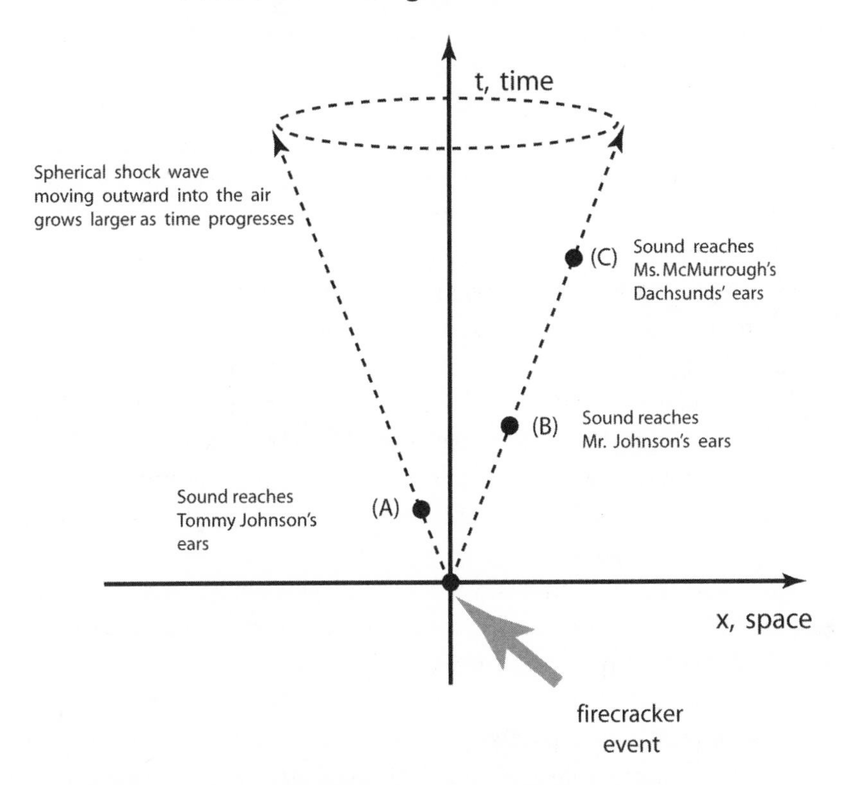

Figure 5.9. Firecracker Sequence of Events. The shock wave of sound emanates outward in space-time from the firecracker event. We see a shock wave that grows larger in radius at later times. This traces out a cone-shaped surface in space-time. Subsequent events (A), (B), and (C) lie on the cone of the shock wave.

And still another instant later, the shock wave has spread farther and reaches the ears of Ms. McMurrough's two dachshunds that are sleeping in her house down the side street behind the hardware store, at event (C). The sound wave continues outward, fading in strength as the compression wave of air expands out into the atmosphere. It leaves behind a startled Tommy Johnson; a concerned Mr. Johnson, who hurries toward the front of the store; and two frantically barking and jumping dachshunds that start climbing up the drapes of Ms. McMurrough's house.

Each subsequent event is defined by the time it takes for the the shock wave from the explosion of the firecracker to reach that particular location in space. The firecracker shock wave is expanding spherically out into the atmosphere, growing larger in diameter but weaker in strength as time progresses. On our plot this shows up as a "cone" in space and time. You need to use a little imagination here, since as we've said, we cannot draw all three dimensions of space (we've depicted the top of the cone by a tilted circle to give the impression of more dimensions of space). At any instant in time after the explosion there is a spherical "wave front" of air compression, the shock wave, i.e., the audible "bang," which we depict by this circle. The distance of the shock-wave-front from the original explosion, x, is just the time from the explosion, t, multiplied by the speed of sound, v_{sound} (that is, distance of "bang" $x = v_{sound}$ times t).

Of course, there's also a flash of light that is emitted by the firecracker explosion. The light travels at c, the speed of light, which is much, much faster than sound, so we can only draw this in a very exaggerated way on our plot. If we also include the light wave in the same plot as the sound wave, it looks like figure 5.10.

Note that the light always reaches the distant observers much sooner than the sound wave. That's why these events are so crowded together near the origin, because the light arrives in such a small time interval, but if you look carefully at figure 5.10 you'll see that these events all happen at the same location in space as the sound events. At event An the light flash reaches Tommy Johnson's eyes. The time interval between event (A) and event (A′) is so short that Tommy hardly notices any time difference between hearing the explosion and seeing it. Mr. Johnson is farther away in the backyard of the hardware store when the direct photons from the firecracker reach him at event (B′), with the sound arriving a tiny instant later at (B). Finally, down the street the dachshunds, though sleeping (well, er, um, dachshunds are never 100 percent asleep), see the flash of light at event (C′) and are alerted, and then a noticeable instant later they hear the boom at event (C), which drives them into dachshund high gear whence they go tearing around Ms. McMurrough's house.

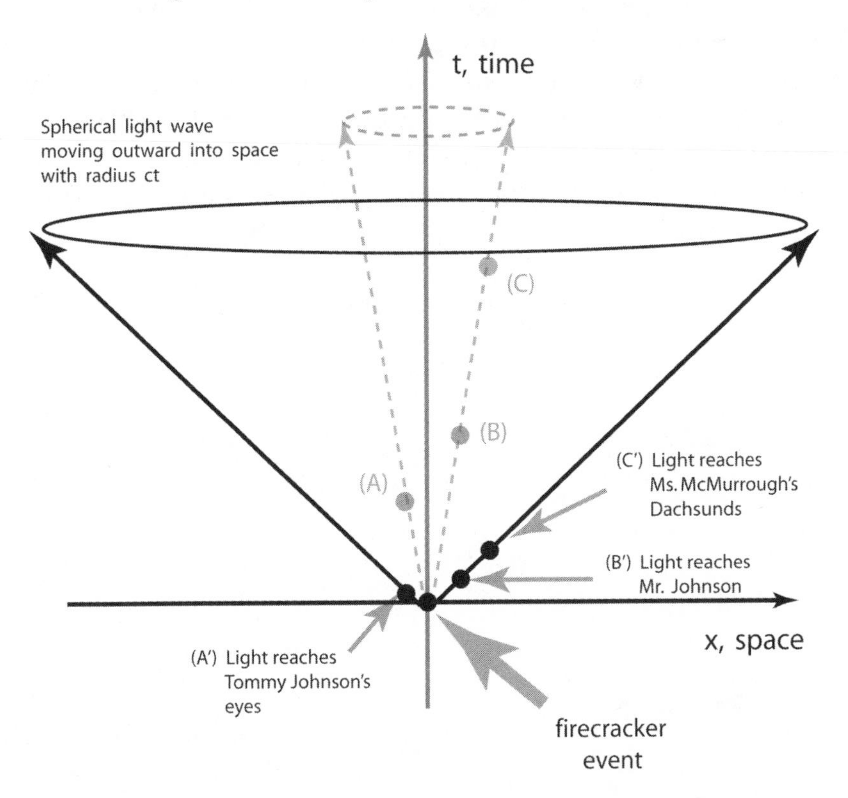

Figure 5.10. Light Sequence of Events. Light also emanates outward in space-time from the firecracker event. This also traces out a cone-shaped surface in space-time. Subsequent events (A´), (B´), and (C´) lie on the light-cone. Note that each of these events occurs at the same space-location, but at much earlier times, than the corresponding events (A), (B), and (C), due to the fact that the speed of light is much greater than the speed of sound.

The very fast light wave continues to propagate outward through space at 186,000 miles per second. Unlike sound, which can only propagate relatively slowly in the air (about 1,000 feet per second) and cannot go into outer space, the light waves from the firecracker explosion continue far, far out and away from the arth. Within a second and a half the light reaches the moon, and could, in principle, be detected with an extremely powerful telescope there. About 38 hours later the light reaches the orbit of Planet Pluto; about 3.8 years later the light reaches the nearest star, ε-Proxima; about 30,000 years later it has transited the diameter of our Milky Way galaxy; in about 13 billion years it reaches the most distant objects we have ever seen in a telescope, the galaxies whose light is now reaching us as they appeared in their embryonic form, just forming after the big bang.

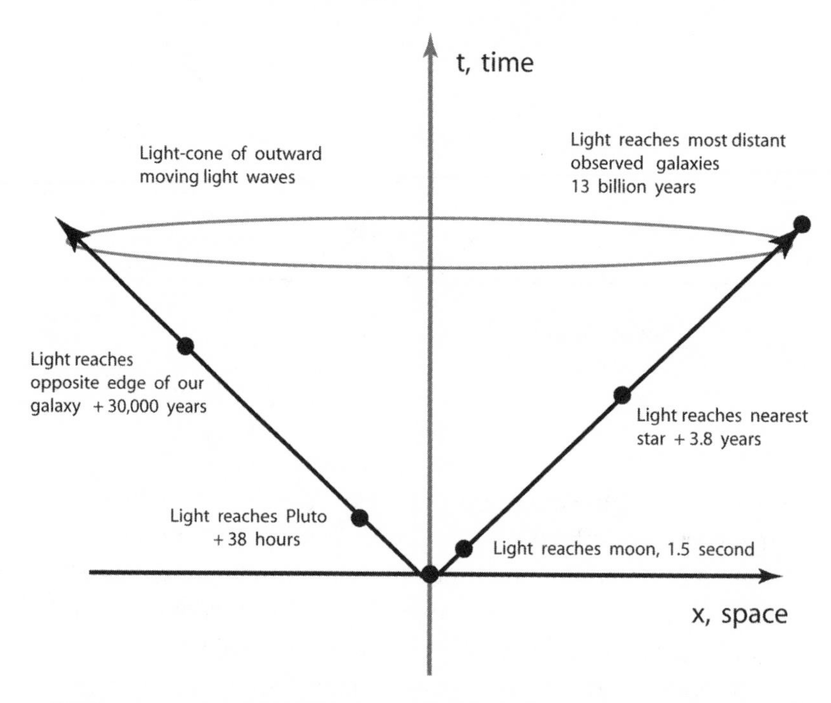

Figure 5.11. Cosmic Sequence of Light Events. Light emanates further into the universe from the firecracker event. At any given time, t (we call it a "time slice"), the radius of the sphere of photons is given by (radius) = c t.

MOTION IN SPACE-TIME

We can use space-time diagrams to represent things that happen in any physical process. For example, we can use a space-time diagram to represent the motion of particles. Light is made of photons that always move at the speed of light. We can trigger the emission of photons by an event at which a flashbulb goes off. Photons then propagate outward in all directions at the speed of light. In space-time the photons are seen to move forward in time at ever-increasing distances from the flashbulb. This traces out a cone in space-time, what we call the "light-cone," spreading out in space and time from the point at which the photons were initially produced.

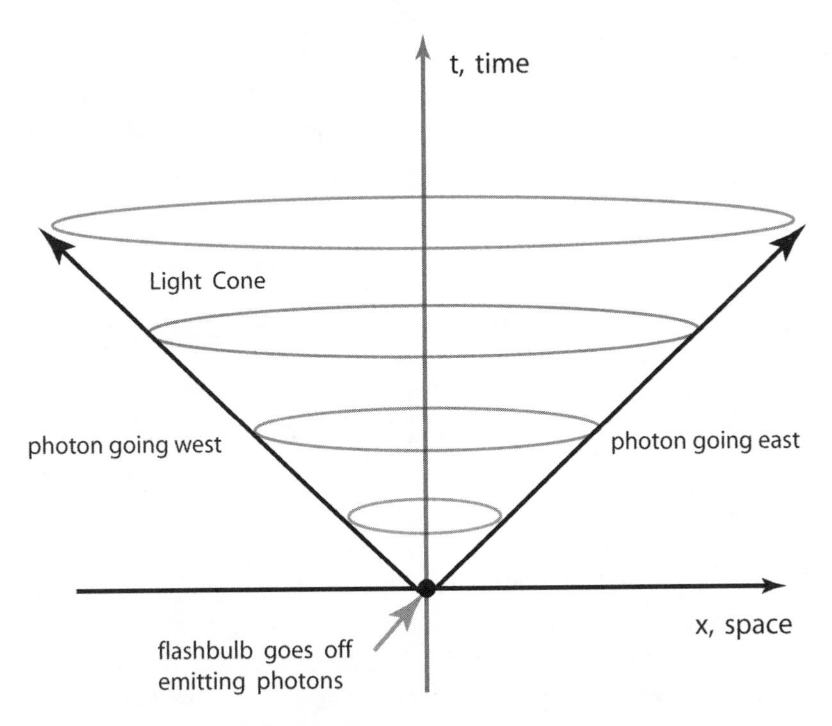

Figure 5.12. The motion of two photons is indicated by the arrows. These individual photon paths lie on the "light-cone," emanating into the future from an event that emitted them at the origin. Photons always travel at the speed of light, c.

By comparison, a single massive particle, like a muon, can in principle travel at any speed, up to nearly the speed of light. A muon, because it has mass, can also sit still. These possibilities are shown in figure 5.13. A muon at rest simply moves, like we all do, forward in time with no progression in space. On the other hand, a very fast muon is one that moves forward in time but also progresses outward in space in some direction.

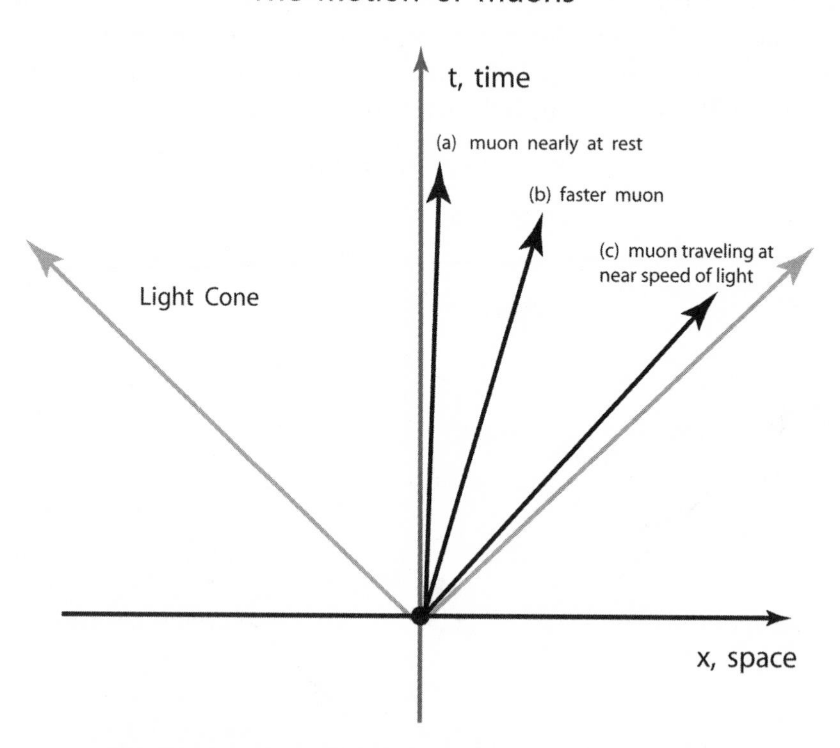

Figure 5.13. Motion of Muons. A muon can have arbitrary velocities less than the speed of light. (a) shows a muon nearly at rest that "moves" forward in time but not in space. (b) shows a faster muon. (c) is an ultra-fast muon traveling at nearly the speed of light.

We can make a muon move very, very fast, but we can never get any massive particle to travel at exactly the speed of light. At the LHC, protons are made to travel at 99.999999 percent of the speed of light. The muon's little sister, the electrons at CERN's LEP synchrotron, were made to travel 99.9999999925 percent of the speed of light. And we have very sensible ideas about someday building a Muon Collider in which we would get muons to travel at 99.99999999 percent the speed of light. But never will an electron or a proton or a muon get to 100 percent of the speed of light. Nature has an ultimate speed limit that is the speed of light. And the existence of mass implies, according to Einstein, that it would take an infinite amount of energy to make any massive particle travel at the speed we call "c."

But suppose, somehow, we could do a master experiment: make a muon have zero mass. We don't know how to do that exactly, but we could get very, very close to this situation experimentally, by making the muon travel as close to the speed of light as possible relative to us. The closer we get the muon to the speed of light, the more and more the muon behaves, as we observe it in our lab, like a massless particle. We can do this with our very powerful future Muon Collider, and as the speed of the muon approaches c, the effects of its mass become undetectable to us. So, what happens to a very, very fast muon as it starts to act like a massless particle?

MUON AT THE SPEED OF LIGHT

From a distance we see a small single-prop airplane flying through the air—it appears to be moving in straight and true line on this clear, sunny day. But what about its propeller? The propeller is moving along the same straight line, but the tips of the propeller are executing a corkscrew-like motion. This is how any spinning object—in this case, the propeller—moves through space and time when it is also in uniform motion.

Recall that the muon has spin. We can never stop a muon from spinning. The spin of a muon is always "up" or "down" along any axis. So we find, upon very careful observation, that a muon (or an electron or a quark) traveling at a high speed moves through space-time like a propeller. The muon, like a corkscrew entering a wine bottle cork, or a drill bit drilling into wood, spins either clockwise or counterclockwise as it progresses through space. This is the "up" or "down" binary nature of the muon's spin combined with motion.

But time is frozen as the muon approaches the speed of light. How can it spin if time is frozen? The way to think of this is that the muon's path through space-time is like that of a corkscrew, either corkscrewing to the right or to the left. The two different corkscrewing paths are the two quantum states of spin, "up" or "down." As the muon approaches the speed of light, these two quantum states become completely independent of one another—the muon has become schizophrenic—it has split into two different personalities altogether as it approaches the speed of light!

This splitting in two of the muon is the consequence of effectively turning off the mass of the muon by approaching the speed of light and by freezing time. The mass of the muon blends the two personalities into one, the usual muon that is heavy and at rest in our lab. But without mass, the muon always travels at the speed of light and then becomes one of two separate and different and independent entities, either a clockwise or counterclockwise rotating corkscrew, or propeller, or drill bit, or whatever metaphor you fancy—it's just one of two different possible massless muons.

If at this point you are starting to feel a bit uneasy, that somehow the elegant simplicity of mass as a mere "quantity of matter" is about to be lost forever, then we suggest that you open a fine bottle of Pinot Noir with a corkscrew wine bottle opener. Now, of course, your bottle opener will turn in one particular way as it penetrates downward into the cork in only one direction. My corkscrew turns clockwise as it descends deeper into the cork.

We have to be precise about what we mean by "clockwise"; that is, we define "clockwise" by looking down from above along the shaft of the corkscrew to the top of the wine bottle. And you can withdraw the corkscrew from the cork by turning it the other way (counterclockwise). And that is an important feature of a drill or a corkscrew—it will rotate one way as it goes in one direction, and the opposite way when it goes the opposite direction! As you contemplate a fresh glass of Pinot Noir, try to figure out if your corkscrew rotates clockwise or counterclockwise as it goes into the cork. To our knowledge, for no particular reason, all corkscrews are manufactured to turn clockwise (looking down from above) as they go into the cork. But there's no reason in principle why there cannot exist a counterclockwise rotating corkscrew. It's just a question of how they were fabricated. Call up the factory and order a dozen counterclockwise corkscrews. Maybe some corkscrews are counterclockwise, while most are clockwise—we're not sure. So, if there are clockwise and counterclockwise corkscrews, these are independent objects like the two pieces of the muon that become separate and independent when we turn off the muon's mass.

CHIRALITY

There's a fancier and more sophisticated way to describe this. We'll assume that you are right-handed (this is unfair to you southpaws, but it is just the way things are defined, so please accept our apology). As you rotate the "clockwise" corkscrew with your right hand by curling your fingers around the knob or handle on the corkscrew, in the manner shown in figure 5.14, you will see that the progression of the screw into the cork is pointed in the direction of your thumb. This is also true for most wood screws or metal screws when you are tightening them with a screwdriver. It's called the "right-hand rule." The right-hand rule states that "the direction of progression of (most) screws is the direction of your thumb as you rotate the screwdriver by curling your fingers of your right hand around the handle."

A Right-Handed Corkscrew

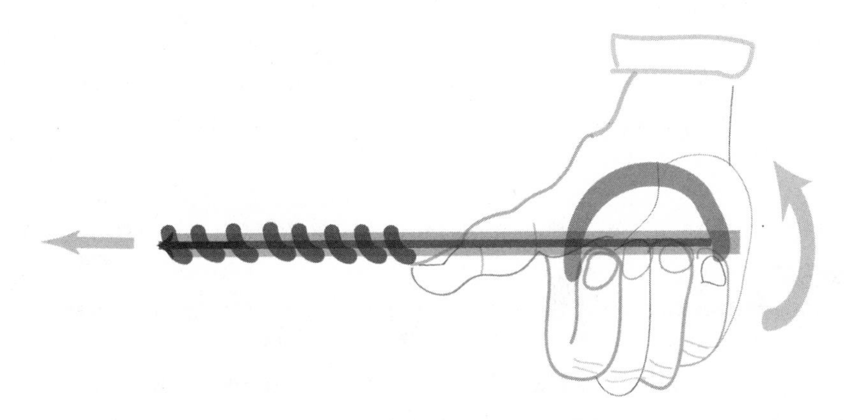

Figure 5.14. Corkscrew. A right-handed corkscrew will progress into the cork in the direction of the thumb of the right hand as the fingers curl around the handle of the screw.

For any rotational motion that is also accompanied by a linear progression, we say the system has "chirality." Our corkscrew in the above example has "right-handed chirality," and we'll call it chirality "R." But, as we said, we can always manufacture a corkscrew that advances into the cork as we rotate counterclockwise. The progression into the bottle as we turn the handle with our left hand would then point in the direction of the left-hand thumb. This is a corkscrew with the opposite chirality, a "left-handed chirality" corkscrew, and we'll call it chirality "L." Likewise, we can have an ordinary wood screw that is "left-handed" and requires rotating the screwdriver in the opposite way to drive the screw into a block of wood.

THE SPACE-TIME PICTURE WITH CHIRALITY

The approximately massless muon, traveling at almost the speed of light progresses through space as much as it is progressing through time, either with chirality L or chirality R.

So now we can depict our massless muon as it travels at the speed of light. It is either a right-handed, R, or a left-handed, L, particle. If the spin of a particle is pointed along the eastern direction as it moves east, then it is R; and if the spin is still pointing east but the particle is moving west, it is L. The R muon state is completely independent of the L state of the muon—the muon has essentially broken apart into two distinct particles, L and R.

Of course, this ambidextrous L and R quality of very fast particles like high-energy muons comes from the quantum phenomenon of spin. But the two spin states of the *resting* or slowly moving muon are easily related: we can simply rotate the muon and one spin (e.g., "up") flips into the other ("down"). And, the *resting muon has no chirality*—it is sitting still, so there is no "progression through space" associated with the spin. But, as the muon travels near at the speed of light, we cannot rotate one chirality state into another anymore (we would have to stop the muon to do this). For a muon traveling east, the two spin states of the muon that were "up" or "down" have now become "spin pointing east" (R chirality) and "spin pointing west" (L chirality) and are now two independent particles. Note that chirality, L or R, is the *combination of the direction of the motion and the direction of the spin*. Chirality involves both of these concepts, linear motion and spin, combined together at the same time.

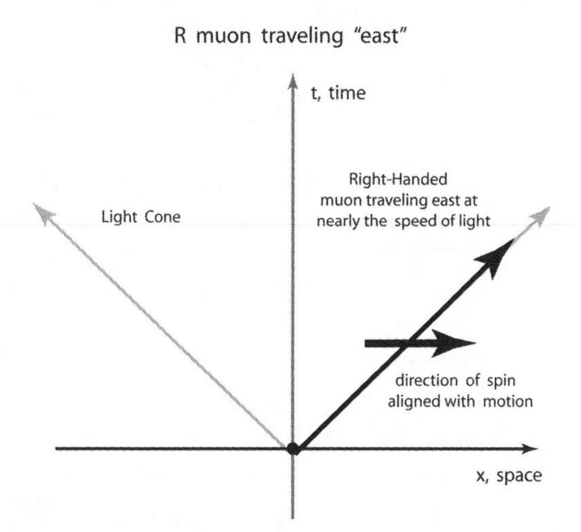

Figures 5.15a–d. Motion of Muon through Space-Time as chiral. Figure 5.15 a. A right-handed particle has its spin (as defined by the right-hand rule) aligned with the direction of motion. As the particle approaches the speed of light we call this "right-handed chirality." Here the particle is moving "east" while the spin is also pointed "east."

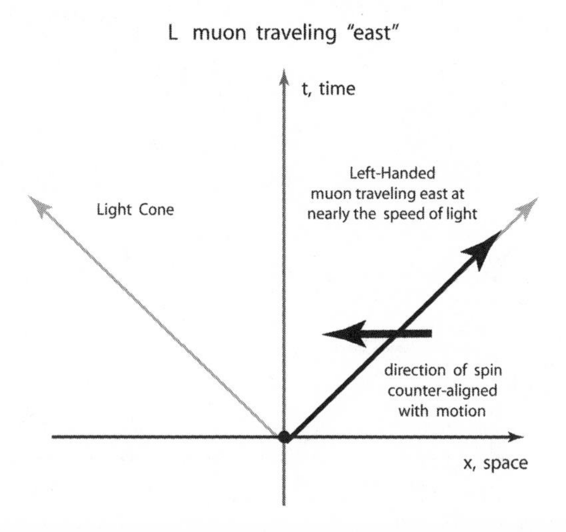

Figure 5.15b. A left-handed particle has its spin (as defined by the right-hand rule) aligned opposite to the direction of motion. As the particle approaches the speed of light we call this "left-handed chirality." Here the particle is moving "east " while the spin is pointed "west."

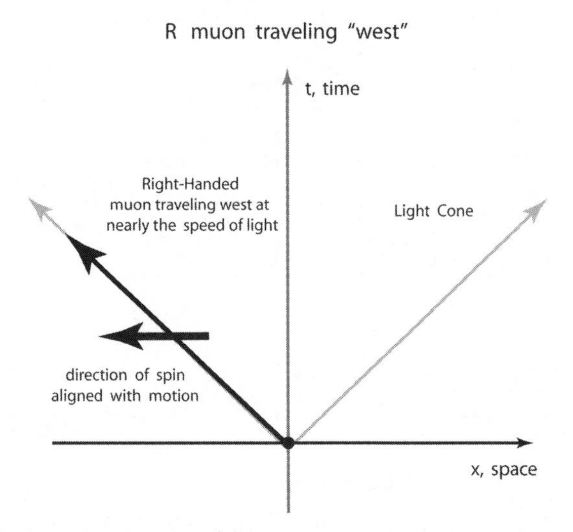

Figure 5.15c. A right-handed particle has its spin (as defined by the right-hand rule) aligned with the direction of motion. Here the particle is moving "west" while the spin is also pointed "west."

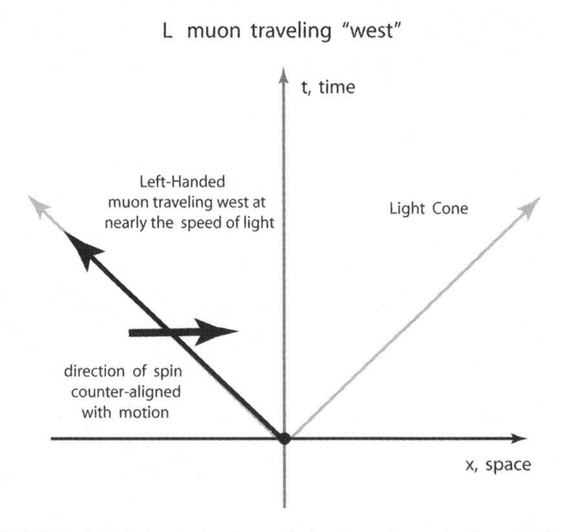

Figure 5.15d. A left-handed particle has its spin (as defined by the right-hand rule) aligned opposite to the direction of motion. Here the particle is moving "west" while the spin is pointed "east."

THE FORCES OF NATURE KNOW ABOUT CHIRALITY

The forces of nature respect certain laws that govern everything. Perhaps one of the most important of these is the law of conservation of energy. The total energy of a system before an interaction occurs is the same as the total energy afterward. Likewise, you are probably familiar with the conservation of momentum, which is why it's hard to stop on a slippery surface, and the conservation of angular momentum, by which gyroscopes always like to point in the same direction and by which it is even possible to ride a bicycle or a motorcycle.

Now we learn a new and somewhat obscure conservation law: it is a stunning fact that all known fundamental (gauge) forces among elementary particles in nature share a special property. They also "conserve chirality."

For example, if a photon interacts with an R particle, the particle will remain R (see figure 5.16a). A photon cannot convert R particles into L (or vice versa). This means that the R particle is really quite independent of the L particle as far as electromagnetism is concerned (so, too, the strong force of quarks, where we replace the photon by the "gluon," and likewise for the weak force, which we'll soon discuss in detail). The interaction strength of the force of electromagnetism, i.e., the electric charge, of an R muon is exactly the same as that of an L muon, even though a photon only makes R go to R, and L go to L.

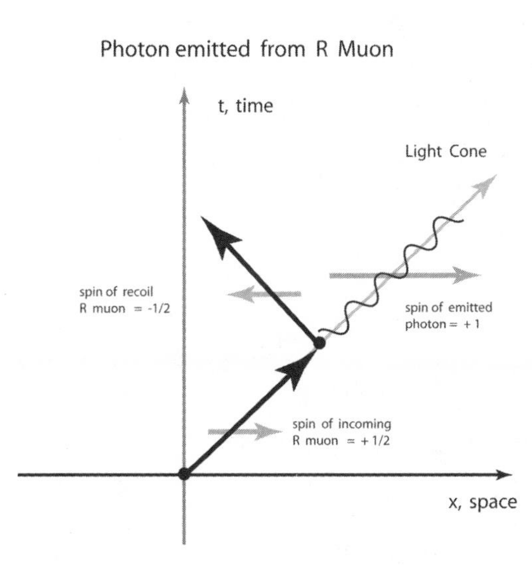

Figure 5.16a. A photon is emitted from an incoming right-handed chirality muon. The muon recoils (changes direction) and its spin simultaneously flips, so that the outgoing muon is also right-handed chirality. Note that the initial spin is +1/2. The emitted photon has spin +1, and the outoing muon has spin −1/2, so the total spin angular momentum is conserved since +1/2 = +1 + (−1/2).

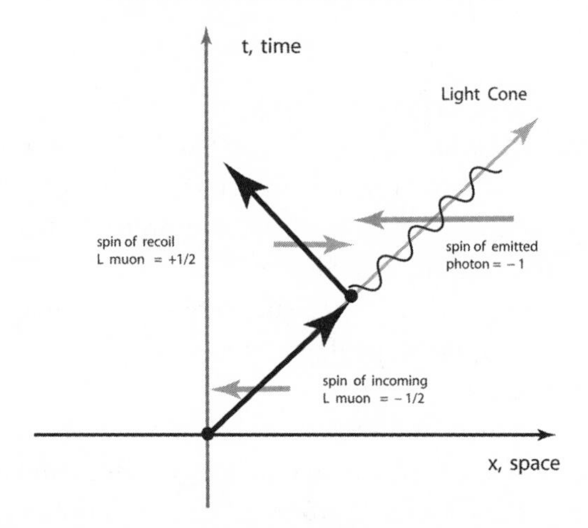

Figure 5.16b. A photon is emitted from an incoming left-handed chirality muon. The muon recoils and its spin flips, so that the outgoing muon is also left-handed chirality. Note that the initial spin is −1/2. The emitted photon has spin -1, and the outoing muon now has spin +1/2, so the total spin angular momentum is conserved since: $-1/2 = -1 + 1/2$.

RESTORING MASS TO THE MUON

Now let's turn the muon mass back on. Of course, our muon always had a mass, but we made its effects unnoticeable by accelerating the muon up to the speed of light. Mass permits a particle, such as our muon, to travel at any speeds less than c, or to sit still at rest.

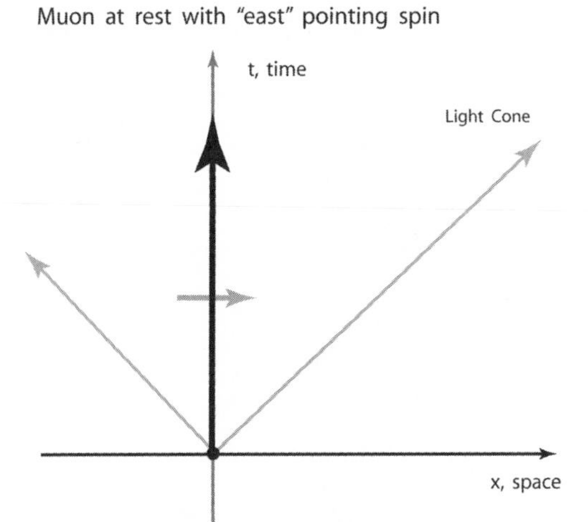

Figure 5.17. A muon at rest has a spin that points in the space direction, but it has no spatial velocity. Therefore, it has no definite chirality. The chirality is meaningful as a symmetry only when the muon is massless, travels at the speed of light, whence the R and L muons become distinct particles. Chirality then becomes conserved.

In figure 5.17 we see a massive muon at rest. It simply moves forward through time, with no progression through space. It has spin, but it has no chirality, because there's no direction in space of velocity to compare the spin direction to. Somehow our two independent chiral states of the muon, L and R, must still be there, but they are now blended together to make a resting muon. How?

Nature does this trick by the effect of mass. But mass in particle physics has a new meaning. *Mass makes an L chiral particle oscillate into a R chiral particle and back again*, without changing the direction of the spin (recall that spin is conserved, the conservation of angular momentum, and the resting muon isn't interacting with anything that can flip the spin around). Mass changes chirality from L to R or vice versa.

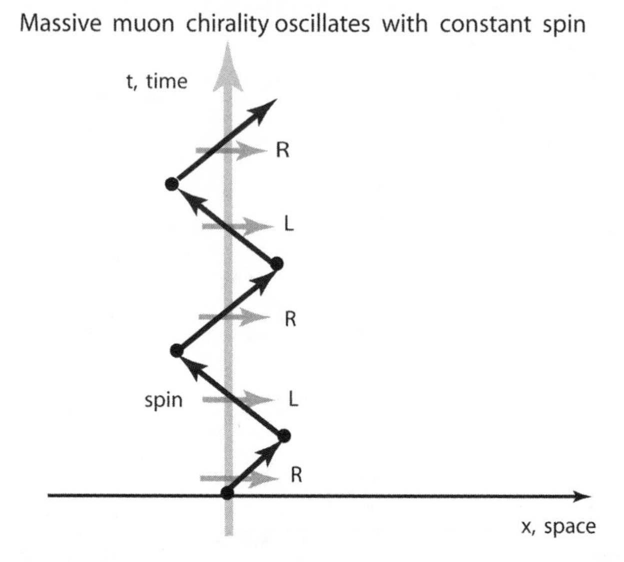

Figure 5.18. At the quantum level, a muon at rest has a definite spin, but it is oscillating rapidly between L and R chirality.

So, the resulting motion of the resting or slow, massive muon, in terms of its two chiral parts, is actually a kind of forced march. The muon goes R-L-R-L-R . . . , to the grunting call of the drill sergeant, called "mass," even though it is sitting still, simply moving forward in time. Each time the muon oscillates, it flips from being an R muon to an L muon back to an R muon again, over and over. The chirality *is not conserved*, that is, it is not always the same for particles that have mass. L and R are blended into one marching particle that hobbles through space and time oscillating between its schizophrenic parts of left- and right-handed chirality.

In this R-L-R-L oscillation the direction of the spin stays the same (the spin is "conserved"), and the electric charge remains the same since L and R muons both have the same electric charges. The R muon and L muon oscillate back and forth between one another, so on average the muon can sit still in space, or move along very slowly with a definite value of its spin in the eastward direction. We have a name for this rapid oscillation of the muon between its two chiralites: we call it "Zitterbewegung."[5]

So the ancient concept of mass is now seen to be richer in the fundamental world of the elementary particles out of which we are built. It now involves "zitter"—the rapid oscillation between L and R. As the muon approaches the speed of light, time becomes frozen, and the muon becomes frozen into either an independent L or R state.

But now let's reconsider the conservation of electric charge. The chirality was conserved by the interaction of the muon with the photon, even though chirality flips L-R-L-R . . . due to mass. But this means that we can make a massive muon (or a massive electron, or a massive proton, or massive quarks inside the proton) out of two different pieces, L and R, *only if the two pieces have the same electric charge*. Electric charge cannot be created out of nothing or made to vanish into nothing. That is, *electric charge must be conserved in nature*: this is a fundamental law of nature.

If the R muon had an electric charge, which we take to be –1 (in some units; this is a standard definition of the electric charge), and if the L muon had, let's say, 0 electric charge, then our R muon could not possibly turn into an L muon because it would be converting charge –1 into charge 0. That would violate the conservation of electric charge symmetry. That is not permitted!

The uniting of L and R parts to make a massive muon works just fine: the electric charge is the same for L and R, so it is conserved. It would appear that there is "no problemo" for a muon, or an electron, or quarks, etc. to oscillate rapidly between L and R states, ergo, to have mass.

HOLD ON A MINUTE!

Hamlet has been known to say that there's more than what we dream of in our philosophy. This is one of the most profound notions in all of Shakespeare, and it captures why physics is such a fascinating subject. Physics is the ultimate philosophy about nature and reality. And whenever we think we've gotten close to understanding it all, Hamlet pops up and reminds us that there is much more.

Let's return to that peculiar radioactive decay of the muon. That guy Lederman and his colleagues showed that the muon decay violates parity. That means it looks different in the mirror world of Alice than it does in our world. But remember, left and right are always swapped in the mirror world—L becomes R in the mirror world and vice versa. Therefore all chiralities are flipped in the mirror. So what is the parity violation effect of Lederman et al. telling us?

When we dissect the muon decay, we find that it is *only when the muon has flipped into its L chiral part that it decays*. The short of it is: the weak interactions *only involve the left-handed L leptons and L quarks* (this also implies that weak interactions only involve the right-handed, R, antileptons and antiquarks). This is the reason why Lederman et. al. observed parity violation in the weak interactions.

The R chiral part of the muon does not feel the weak force. By itself, the R muon would be stable. But the muon is not stable because, as we have just seen, it always oscillates back and forth between the two chiralities, R and L, due to its mass. So the muon at some point will always flip into its L chiral part, and then it can decay.

The weak force "knows" something about the L particles that it doesn't share with the R particles. The weak force involves only L but not R. *But the weak force also has a conserved charge, just like electromagnetism.* Only the L muon has this weak charge—the R muon has no weak charge. Sometimes we say that the R particles are "sterile" under the weak force.

Alas, therefore, the L-R-L-R . . . oscillation of a massive muon ought not to be allowed! It would violate the conservation of the weak charges. Yet it does happen! How can we, therefore, combine two totally different things together, L muons (with weak charges) and R muons (with no weak charges) to make the whole muon that sits at rest on a table? That is, how can the muon, or the electron, or the quarks, etc. have mass?

Please reread the last few paragraphs. It's really simple logic, but there are a few things to keep track of. But we're almost starting to glimpse the Higgs boson. It's a short climb to get there. Do you like rainbow trout?

CHAPTER 6

THE WEAK INTERACTIONS AND THE HIGGS BOSON

We're going to go fishing for rainbow trout. There's a beautiful mountain lake, high in the Rockies. It's a bit of climb to get there, but it's well worth it. Not only will the exercise be good for us and the fish delectable, but we'll see the most beautiful pinnacles in nature. Are you ready? Got your hiking boots and your fishing rod? Let's go.

As we have seen, the muon decomposes into two independent components when it travels nearly at the speed of light, where the effect of mass becomes irrelevant. It's as though we have banished the muon mass altogether. One of these pieces is R (the spin is pointing in the direction of motion) and the other is L (the spin pointing opposite to the direction of motion). The mass of the muon creates the union of these two pieces into one. A slowly moving muon (much less than the speed of light and perhaps at rest) marches through space-time, oscillating between L and R like the drone of a military marching drill, . . . L-R-L-R. . . . At the speed of light, time becomes frozen for our muon, the oscillation stops, and the muon can become one of either the pure L or pure R state.

All matter particles, the electron, the tau, the quarks, even the lowly neutrinos that have miniscule masses, oscillate in this fashion. Mass, which was the measure of the "quantity of matter," inherited from antiquity, now has a deeper meaning in particle physics. L and R are like the "atoms" of the phenomenon of mass. Only in the utopian world, in which all particles have zero mass and always travel at the speed of light, is the union of L and R broken apart such that particles decompose into their two completely independent entities. In the massless world the L and R components of every quark and lepton take on their own unique identities.

But how does nature perform this grand marriage of L and R that gives us mass, the things we see in the world around us, the world of things moving slowly or at rest, having time to think and experience?

In a simple world in which there are only muons interacting through electromagnetism (this is called "quantum electrodynamics," or "QED" for short) we can easily perform this happy marriage and still respect the vaulted law of the conservation of electric charge. That is, there is nothing special about mass in QED. This can be done because the R muon has the same electric charge as the L one. For the muon, as well as for the electron and all electrically charged matter particles, when we include the effects of electromagnetism, there is a perfect symmetry between L and R—"parity" becomes a symmetry and L and R are otherwise indistinguishable—they have the same electric charges.

From Alice's point of view, she might encounter the L and R muons on the table in her parlor and then see the mirror reflection of L and R, but by observing only their electric charges, she would see no difference between the mirror L and R in the looking-glass house, because L and R have the same electric charges. Theoretical physicists introduced the mass of the muon, or of the electron, into their equations "by hand" in the theory of QED—there was no need for a Higgs boson in electromagnetic theory.

But, as we described in chapter 3, parity was discovered experimentally not to be a symmetry when the weak force is involved. The experiment of Lederman and his colleagues in 1957 demonstrated for the first time that parity was violated both in the weak interaction decay of the pions of Yukawa and of the muon itself—pretty good for a weekend's work! It was astounding news—the weak processes are not invariant under the parity—the looking-glass house through the mirror has different laws of physics than in Alice's parlor on our side of the mirror. This means that if Alice looks closely enough at the L muon, she will see something quite different than for the R muon.

THE WEAK INTERACTIONS

It's been almost 70 years since Enrico Fermi wrote down the first descriptive quantum theory of the "weak interactions." At that time, these weak forces were backstage, the feeble forces seen at work in nuclear processes such as beta decay. Only later was it understood that they are critical to the burning of the sun and provide the gunpowder of nature's largest explosion since the big bang, the supernova. Supernovas make the heavy elements found in the universe and especially here on Earth. Without the weak force in nature we wouldn't be here.

So, we're going to fast-forward through history. The weak forces were later found to be very similar in structure to electromagnetism. Like the photon of electromagnetism—the particle that jumps back and forth between charged particles and creates electric and magnetic forces—the weak interactions also involve new particles, called W^+, W^-, and Z^0. These are similar to the photon, and the quantum jumping of W^+, W^-, and Z^0 between matter particles causes the weak forces.

Just as the particles that "feel" electromagnetism all have electric charges, the matter particles that feel the weak force have "weak charges." Unlike the photon, however, the W^+, W^-, and Z^0 are very heavy particles, and this suppresses their jumping back and forth between particles that have weak charges, so the weak forces become very weak. Initially, the W^+, W^-, and Z^0 were only theoretical discoveries, but they ultimately defined a big part of the architecture of what we now call the Standard Model. These developments were led by theorists Sheldon Glashow, Abdus Salam, and Steven Weinberg, and the theory was perfected into a workable quantum theory by Gerard 't Hooft and Martinus Veltman, all of whom shared well-deserved Nobel Prizes for their heroic effort.

The weak interactions are welded, or "unified," with electromagnetism in the Standard Model. In the utopian world in which we can turn off all the masses of all particles, the W^+, W^-, and Z^0 would also become massless and are essentially indistinguishable from the photon, γ. The "symmetry" of the Standard Model is precisely the idea that, when they are all massless, these four particles can be viewed as one "uber-particle" with four parts (technically it's 3 + 1 parts, but we'll not get into the delicacies of this distinction). We now know how to effectively turn off mass: just make things travel as closely as you can at the speed of light. It's a little hard to do, but that's almost what happens at the LHC when these heavy particles are produced. And, indeed, we see in our experiments that at ultra-high energies, or for extreme short distances and short time scales, the symmetry of the Standard Model works down to every last detail. There are many predictions of the Standard Model that have been tested experimentally, and we haven't found a single glitch yet.

Developing the Standard Model constituted a revolution in particle physics that occurred in the early 1970s, at about the time quarks, the tiny particles that make up the proton and neutron and pion, were first glimpsed in experiments. This was the decade when it became both theoretically and experimentally established that all forces in nature are governed by the overriding symmetry principle, called *gauge symmetry*. This was known to govern electromagnetism and gravity and could now be extended to the weak interactions and the strong force among the quarks.

HOW DO THE WEAK INTERACTIONS WORK?

Let's return to our "space-time" diagrams. In figure 6.20 we show a greatly magnified view of how a muon decays, as Fermi would have described it in his primitive theory of 1935. The heavy muon is at rest and it moves forward in time. Suddenly, it disintegrates into a low-mass electron and two neutrinos. We write symbolically for this process:

$$\mu^- \rightarrow \mathbf{e}^- + \nu_\mu + \mathbf{anti}{-}\nu_\mathbf{e}$$

In Fermi's day the neutrinos were considered to be the same species of particle (one is a particle, the other an antiparticle), and the overthrow of parity had yet to be discovered.

Much was learned since Fermi's original paper in 1935 up to the time of the Standard Model revolution in the early 1970s. Today we know (again, thanks to Leon and his friends and their work on a different experiment, which won Leon the Nobel Prize) that there are actually two different kinds, or "flavors," of neutrinos involved here. One of these is associated with the electron and is called the "electron-neutrino." Another is associated with the muon and is called the "muon-neutrino"; today we know of a third neutrino associated with the τ lepton, called the and τ-neutrino.[1] But the main thing we want to examine is precisely how this process occurs at extremely short distances. In fact, the weak interaction of the muon that produces its decay is an "indirect process," where the W boson is only evident at an extremely short distance over a miniscule period of time. To see this we need to crank up the magnifying power of our microscope by a quite a bit.

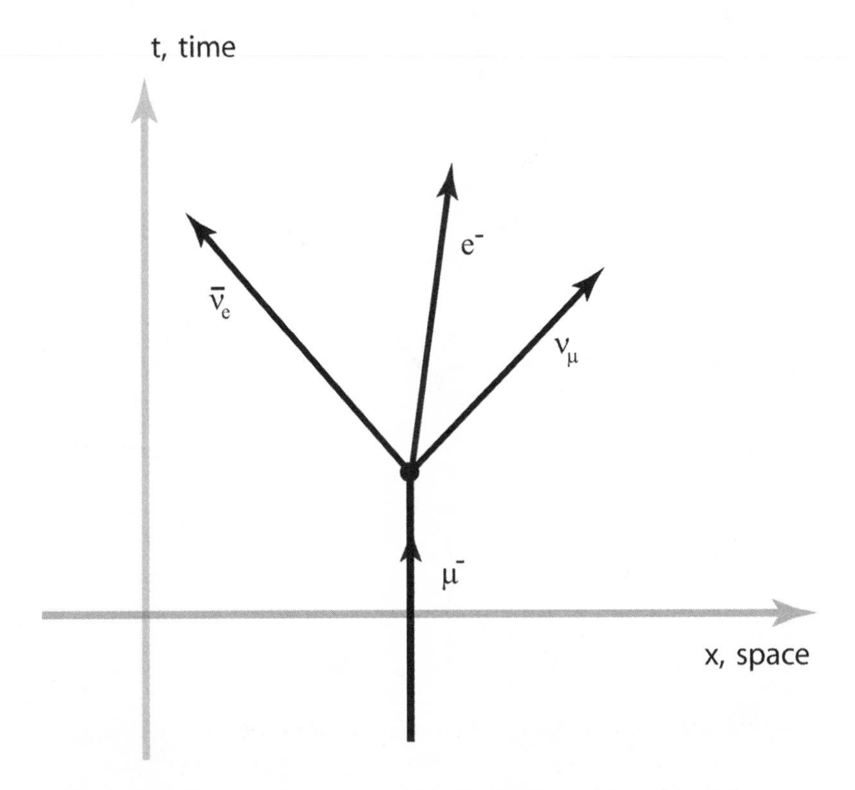

Figure 6.19. Fermi's Theory of Muon Decay. Muon decay is depicted in the Fermi theory of weak interactions, ca. 1935. Here we see the process $\mu^- \rightarrow e^- + \nu_\mu$ + anti-ν_e. Using the measured lifetime of the muon of about 2 millionths of a second, we can calibrate the strength of the weak interactions, and we can infer the "energy scale of the weak interactions" to be about 175 GeV. This turns out to correspond to the strength of the Higgs field in the vacuum.

In figure 6.20 we show how the muon decays as it would be seen at the magnification of the Fermilab Tevatron or the LHC (this isn't exactly how it's done, but the metaphor holds; processes exactly like this were seen in top quark decay at the Tevatron when the top quark was discovered in the mid-1990s and are now "bread-and-butter" physics at the LHC).

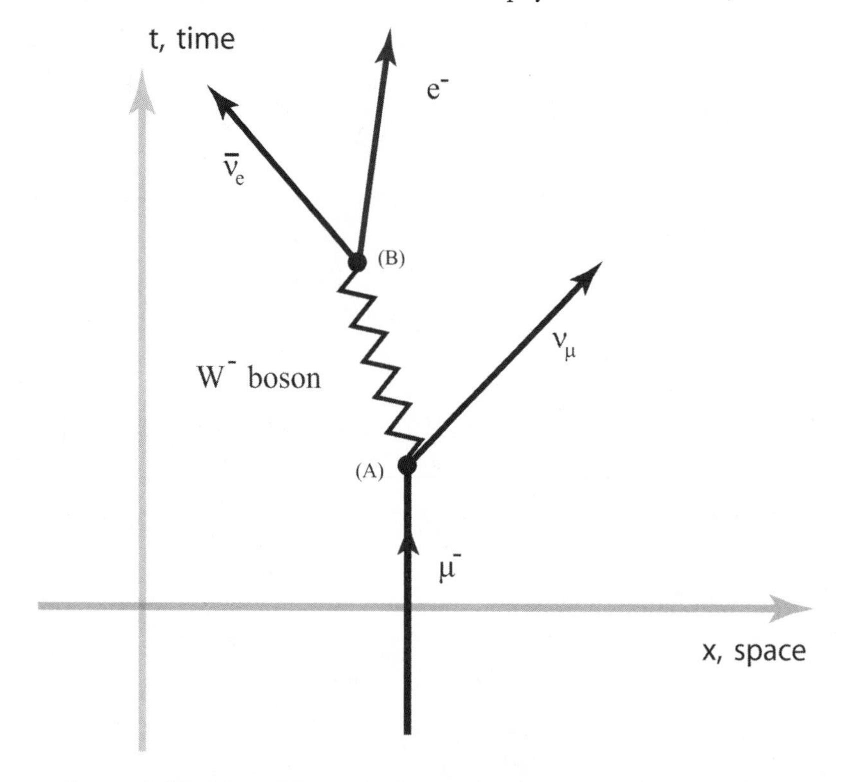

Figure 6.20: Muon Decay in the Standard Model. We examine the muon decay under a powerful microscope, which shows how it appears in reality as described by the Standard Model. We see in the process $\mu^- \rightarrow e^- + \nu_\mu$ + anti-ν_e that a muon converts into a W boson and a neutrino. There is insufficient energy for the W boson to become a real particle in this process, so this only occurs as a quantum fluctuation, for a brief instant of time, as allowed by Heisenberg's uncertainty principle. The W boson instantly converts to the electron and antielectron neutrino, $W^- \rightarrow e^-$ + anti-ν_e. This is a suppressed quantum fluctuation causing the weak interactions to be "weak." By particle physics standards, the muon lifetime of 2 millionths of a second is a very long time.

Here we see a sequence of events. We start with our massive muon at rest. Then at event (A) the muon converts to a muon-neutrino and a W^- particle. Note that the negative electric charge of the initial muon has gone to make the negatively charged W^-, so electric charge is conserved, as it must be. The W^- then instantly converts into the electron and an antielectron neutrino at event (B). Again, electric charge is conserved.

"Wait a minute," says Graham, "aren't you swindling us with this small mass, muon particle converting into the monster heavy W^-? This must grossly violate the conservation of energy!" Indeed, the W^- boson has a mass that is almost a thousand times greater than the muon (the mass of the muon is only 0.105 GeV, while the W^- boson has a mass of 80.4 GeV). Graham is right! There is no way that a muon can convert to a neutrino plus the ultra-heavy W^- boson and conserve energy. What is happening here? Why does an "indirect process" exist with an ultra-heavy W^- boson existing for only a fleeting instant of time?

This is an example of one of the great wonders and chestnuts of quantum theory, called the *Heisenberg's uncertainty principle.* The time interval between the creation for the W^- from the muon, and W^- converting into the electron is extremely short (about 0.0000000000000000000000001 seconds, or 10^{-25} seconds). Heisenberg tells us that, as a consequence of quantum theory, energy is a fundamentally uncertain quantity during extremely short time scales. In fact, he tells us exactly by how much.[2] For that miniscule amount of time, the amount by which the energy is uncertain is equivalent to the mass of the W^- boson in Einstein's formula $E = mc^2$. Therefore, the uncertainty principle allows the W^- to exist, but only for a tiny instant of time. This is called a "quantum fluctuation."

However, this requisite, large "quantum fluctuation of certainty of the energy" needed to momentarily evade energy conservation causes the resulting process to be very improbable—it is a "rare quantum fluctuation." This is why the overall process is a "weak interaction." In fact, the amount by which the overall decay of the muon is suppressed, since it requires a big quantum fluctuation in energy, is a factor of about one trillionth (or 10^{-12}) compared to what would happen if the W^- could be replaced by a massless particle like a photon, which would require no quantum fluctuation in energy at all (the photon, however, cannot convert a muon into its neutrino since it must conserve electric charge and can only convert a muon into a muon).

The only way the muon can decay is through this highly suppressed process involving the heavy W^- boson that converts a muon into a muon-neutrino.

But let us examine the "conversion" of the muon into a neutrino plus a W^- boson in still greater detail. Recall that the world of Alice through the looking glass is fundamentally different than ours because of the violation of parity. If we examine figure 6.21 we see why. The muon comes marching in, L-R-L-R- . . . as all massive particles do, oscillating between L and R. At some instant, while the muon has oscillated into an L particle, it can convert into the W^- boson and the neutrino. The neutrino is effectively massless (it has a mass that is less than 0.00000001 times the muon mass, so it behaves like a massless particle), and an L muon-neutrino emerges, almost on the light cone. The point is that *W^- boson interacts with the L muon and with the L neutrino.*

How do we know that W bosons interact only with L particles? Go back and reread Leon's account of the discovery of parity violation (chapter 3). It was observed that muons decayed (see figures 6.19–20) in such a way that the electron coming out moves in the general direction of the spin of the muon (getting the details of this right is a little tricky). This also means that a direction in space is associated with a spin, and that always implies a preferred chirality, so the decay process violates parity[3].

This is the mysterious source of the parity violation in the weak interactions. It is connected intimately to mass and to the fact that every particle has two inner L and R chirality components. Only the L part of a quark or a lepton can convert to a W boson (and it's reversed for antimatter; only R antiparticles can convert to a W). Alice in the looking-glass house would see L and R swapped. So, she would see that the R muon is converting to the W boson, and her neutrino would be R, with its spin aligned with the direction of motion. But *that is a different world than ours.*

AYE, HERE'S THE RUB

So we're almost there. We've climbed a long way, but the beautiful mountain lake full of rainbow trout is still another hundred feet up. It's only a short climb from here. Take a breath, a swig of water, and let's continue. The first purple peak will soon come into view. We are about to see why a Higgs boson must exist.

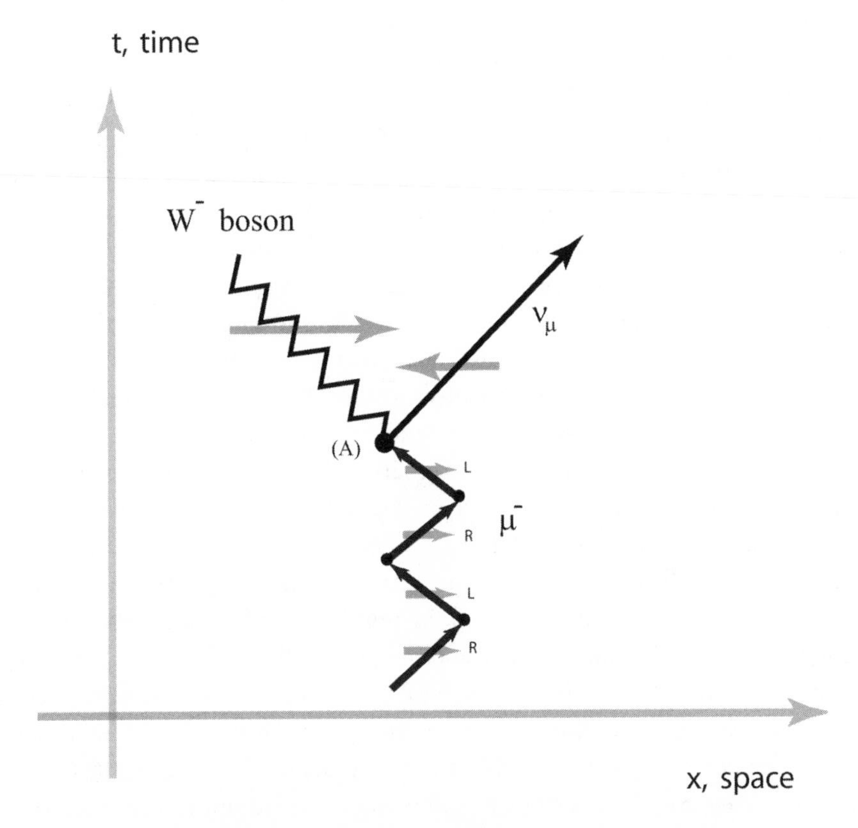

Figure 6.21. Chirality in Muon Decay. We examine figure 6.20 more closely to reveal that only the left-handed muon (and left-handed neutrino) converts to the W boson: $\mu^- \rightarrow$ anti-ν_μ + W^-. Likewise, only the left-handed electron (and right-handed antielectron neutrino) is produced by the process $W^- \rightarrow e^-$ + anti-ν_e. The weak interactions involve only left-handed particles (and their right-handed antiparticles).

The W^+ and W^- bosons are like photons. They couple to the L muon (R anti-muon) and L neutrino (R antineutrino) with a "weak charge." In order to unify the W's with the photon and with the Z boson in perfect utopian symmetry, it is necessary that this charge be very much like the electric charge—weak charge must be conserved. In a more erudite manner of speech: "the defining gauge symmetry principle of the weak interactions is the conservation of weak charge, just like the conservation of electric charge defines electromagnetism."

Why must charges be conserved? What goes wrong if they are not? This is a deep question and has to do with the remarkable principle of "gauge symmetry" that underlies electromagnetism. It's actually implicit in the fancy name "electro-magnet-ism." Essentially, it implies that all we can ever observe in nature about photons are electric and magnetic fields. Electric fields accelerate electrons, imparting energy to them, while magnetic fields bend their trajectories into circular motion. But the photon is actually neither an electric or magnetic field: it is something more basic. The photon is a wave of something called a *gauge field*. The gauge field *cannot be directly observed*. But gauge fields can readily produce electric and magnetic fields.[4] Only the electric and magnetic fields produced by a gauge field can be observed. But if we observe an electric or magnetic field we cannot reconstruct exactly what underlying gauge field made it. And there are nonzero gauge fields that produce no observable electric or magnetic fields.

Katherine exclaims: "So, an infinite number of different possible gauge fields can make the same electric and magnetic fields?" Yes. And that is *the* symmetry. Any two apparently different gauge fields that make equivalent electric and magnetic fields we say are "(gauge) equivalent" to one another. It's like a perfect wine bottle with no label—rotate the bottle about its axis of symmetry, and the wine bottle is now in a different position, but it looks exactly equivalent to the position we started with. We say the two positions of the wine bottle are "rotationally equivalent" to one another. Katherine: "Well, OK, that makes sense, I suppose. But in order to make it work for the wine bottle you had to make sure there was no label—no marks that allow you to tell you rotated the bottle. What's the analogy of that to gauge symmetry?" The short answer is that in order for gauge symmetry to work, the electric charge must be conserved. The total electric charge you start with must be the same as the electric charge you end up with (See our book *Symmetry and the Beautiful Universe* [Amherst, NY: Prometheus Books, 2007] for a much more detailed discussion of gauge theories).

It also turns out that the gauge symmetry principle is intimately related to quantum theory. Without the quantum waves that describe electrons and other charged particles, together with photons, the gauge symmetry seems awkward—there's no electron wave to "transform" under the gauge symmetry—there's no "representation" of the gauge transformation, like the wine bottle that rotates when we do a rotation transformation in space. It was as if electromagnetic theory, created by wizards in the nineteenth century, was waiting and begging for the quantum theory to come along and make it whole. And, if you try to modify electromagnetism to make the gauge fields directly observable, the whole structure of the quantum theory breaks down and becomes a heap of rubble. The key to cloaking the gauge field—making it unobservable while the derived magnetic and electric fields are observable—is the conservation of electric charge. But now we encounter the weak interactions with the three new gauge bosons, W^+ W^- and Z^0 and all particles now have weak charges. Again, as in electromagnetism, we find that the weak charge, like the electric charge, must also be conserved.

The weak charge of the L muon is −1 (in some units). But the parity violations experiment tells us that the weak charge of the R muon is zero. *L particles have weak charge while R particles do not.* So now the marching L-R-L-R-L of a massive muon (or electron or top quark—any other matter particle will do) creates a problem. As L turns into R, the weak charge of the muon changes from −1 to 0. *The weak charge is evidently not conserved for a massive muon sitting at rest, minding its own business, and oscillating between L and R.* Mass breaks the vaulted gauge symmetry, which we have now extended to include the W^+, W^-, and Z^0 bosons. But without conserved charge, *the gauge theory collapses into charred remains.*

Please reread the previous paragraph. This is what directly leads us to the Higgs boson.

The original paper of Sheldon Glashow on the Standard Model of electromagnetic and weak interactions defined the basic structure of the weak and electromagnetic gauge symmetries and introduced the W^+, W^-, Z^0.[5] However, Glashow needed masses to explain the physical world, so he "put them in by hand," knowing this was a serious problem but thinking a solution would come later. This, therefore, was not yet a mathematically complete theory, and it wasn't understood if it could ever be made compatible with the principles of quantum mechanics at the time.

In 1967 Steven Weinberg wrote what has become an iconic paper of the Standard Model, "A Model of Leptons."[6] He took Glashow's theory and proposed a clever remedy to the problem of mass, focusing only on the electron and its neutrino. He was inspired by the paper of Peter Higgs, but had to refine Higgs's idea to make it work. In the end, Weinberg had engineered a kind of "superconductor" that made the W^+, W^-, and Z^0 bosons heavy, while the photon remained massless. He also showed how the masses of all the matter particles, electrons, muons, the top quark, even ultimately neutrinos, could be explained. Without Weinberg's idea it would not be possible to have a consistent theory of mass for any of the elementary particles.

Even after Weinberg launched his paper, many people had reservations about whether the theory really was truly mathematically consistent, and the idea didn't catch on immediately. It took the super heroic efforts of Gerard 't Hooft and Martinus Veltman to show that it was indeed a workable and useful theory, and to show us how to use it correctly.[7] This opened a scientific discovery floodgate, and the "gauge theory revolution" began. A number of key refinements and major extensions were required to accommodate the quarks and the strong interactions. The rest was up to the theorists to compute the various predictions for physical processes, and to the experimentalists to measure these things and test the theory. It has proved to be a stunning success.

ENTER THE HIGGS BOSON

Recall that our problem is to make a massive muon march, oscillating L-R-L-R-L-R, even though each time L changes to R the weak charge changes from –1 to 0. How do we do it? As the muon simply sits at rest it is rapidly "oscillating" L-R-L-R . . . its weak charge is also oscillating: –1 0 –1 0. . . . The weak charge is flickering on-off-on-off. The mass of the muon seems to destroy the neat conservation law of the weak charge. The Standard Model must either be wrong—or something new is happening to rescue it.

Graham: "So, let me ask a simpler question: Is there any way, or any process, even if it's only theoretical gibberish, in which an L muon can convert to an R muon and still conserve the weak charge?"

In fact . . . yes! There is . . . and this is the key! If we introduce a new kind of boson *that has the same weak charge as the L muon*, a boson that also has weak charge –1, then the L muon could "convert" to such a boson, plus an R muon. When the initial L muon has weak charge −1 and converts to the R muon with weak charge 0, plus the new boson that has weak charge −1, the overall weak charge remains the same since −1 = 0 + (−1). The required space-time picture is shown in figure 6.22.

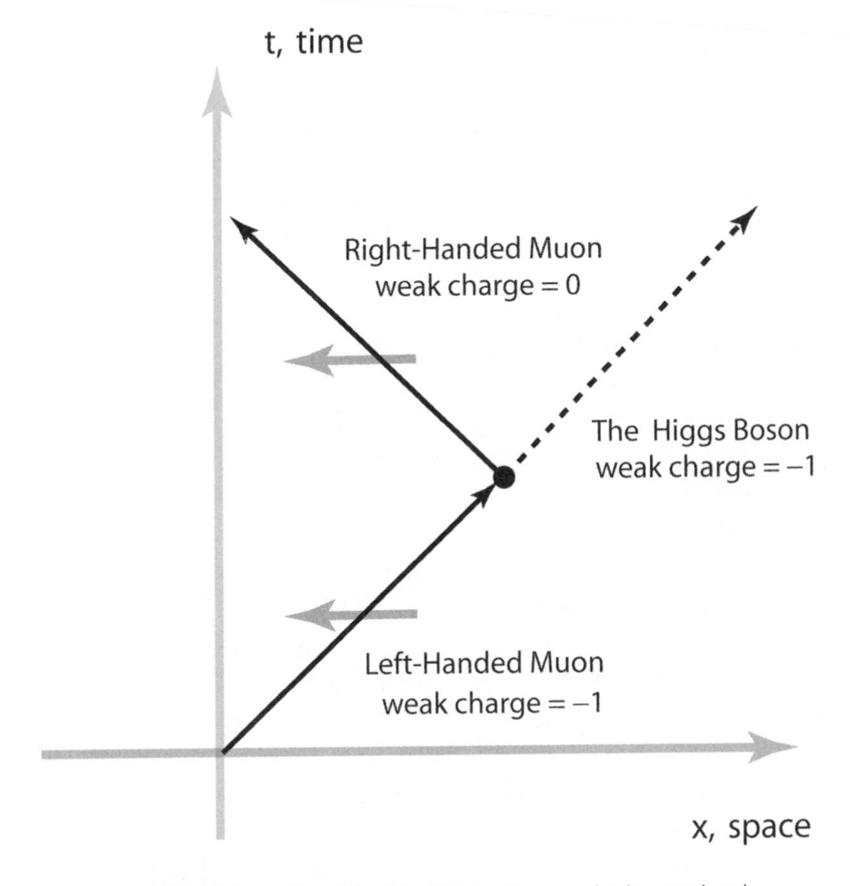

Figure 6.22. Muon Coupled to Higgs Boson. An incoming L muon with weak charge -1 converts to a recoil R muon with weak charge 0, plus the Higgs boson with weak charge -1 and spin 0. The process has a "coupling strength" g_μ. It unites the L muon with the R muon. The Higgs boson must carry a weak charge equal to that of the L muon, so the weak charge is conserved. The Higgs boson must also have spin 0 to conserve spin angular momentum.

The properties of our new boson are completely dictated by this process (this is explained in detail in the caption of figure 6.22). Since the L muon is converting to an R muon, maintaining a constant spin direction, the new boson must have zero spin. Note that since the L muon and R muon have the same electric charge, −1, the new boson must therefore have zero electric charge. And the key to the whole game, since the incoming L muon has weak charge −1 and the outgoing R muon has weak charge 0 is that the new boson must have weak charge −1. Voilà! We have completely spelled out the properties of the new boson. We have done exactly what Weinberg did in his 1967 paper. We have introduced a new theoretical particle, a spin-0 boson, that carries weak charge. We call this the Higgs boson.

BUT WHAT ABOUT MASS?

We're almost there. I can just taste those rainbow trout that we'll soon be catching at the mountain lake once we put our fishing gear into action. It's just another 50 feet . . . take a deep breath. The grandest peak of all will soon come into view.

What do we mean by "boson"? For reasons that are deep and profound and have to do with quantum fields and relativity, there is a remarkable difference between matter fields, like electrons and muons and quarks, etc., and things we call the force carriers, the "bosons." Matter particles are called "fermions" after Enrico Fermi. For no reason connected to their namesake, they are recluses and like to keep away from other fermions (Fermi was quite outgoing and personable).

Fermions like to avoid one another. We can never get two fermions into the same quantum state—they are forbidden from doing this. This is a deep principle and has to do with the weird quantum attribute of spin.[8] This property was discovered in the process of understanding atoms, and without it there would be no chemistry—all atoms would collapse down into different forms of the chemically inert helium—the universe would forever be one big gas bag of non-interacting helium-like atoms.

Bosons are named after the Indian physicist Satyendra Nath Bose, who was a friend of Einstein.[9] Bosons are gregarious particles (unlike their namesake, who was kind of shy). Bosons love hot tubs. They all pile into the same quantum state together whenever they can. In fact, whenever bosons start to pile into the same state, one of them yells "Party's on," and pretty soon a gazillion bosons end up in the same state. You've seen this phenomenon in the dramatic instance of a laser beam, where many photons, which are each bosons, pile into exactly the same state of motion with the exact same frequency and wavelength of light, making a mysterious and intense beam of light.

But there's a more mundane example. Any old electric or magnetic field is just a very large and indefinite number of photons dancing around in a small set of quantum states. This is called a "classical coherent state." The quantum particle aspect of this becomes so blended that all we see is a large macroscopic wavelike field, and it is described by Maxwell's classical equations of electromagnetism. The blending into a coherent or "collective" state masks the quantum nature of the photons that make up the field. The same is true of the radio waves delivering text messages to your iPhone®. These are large aggregates of photons that behave "collectively" like one big field, doing exactly what bosons love to do.

THE HIGGS VACUUM

Weinberg realized that this bosonic "piling on" into one big collective state could also happen to the Higgs boson. We need only create an enormous "Higgs field" that permeates the entire universe. This field is similar to the magnetic field of the earth, in that it is composed of the Higgs bosons acting collectively, while magnetic fields are composed of photons acting collectively. However, magnetic fields have a well-defined direction in space—defined by the direction the compass needle points—we call a magnetic field a "vector field." The Higgs field, on the other hand, just has a value—measured in energy—it has no direction in space. We call it a "scalar field."

Why would the universe have such a field? There is again a clever idea that the particle theorists borrowed from the study of materials—in this case the phenomenon of ferromagnetism. An iron magnet will spontaneously magnetize when it is cooled. The magnetic field seems to pop out of nowhere, but it is actually coming from the trillions and trillions of atoms inside the iron magnet. Each atom has a spin and is a little magnet itself.

For iron, when it is cooled, these atomic spins all line up and point in the same direction. This creates a large magnetic field.[10]

While magnets are complicated in detail, we can understand iron in a simple way through a plot. The point is that the energy of the iron is reduced the more the atoms' spins are aligned. This comes from the complex interactions of the atoms with each other in the iron material. If we have misaligned atoms, or random atomic spins that have no net alignment, we have a state of higher energy. The state in which all atomic spins are aligned has much less energy, and for some particular amount of alignment, or "magnetization," we get the lowest energy state. The magnetic field then appears *spontaneously*.

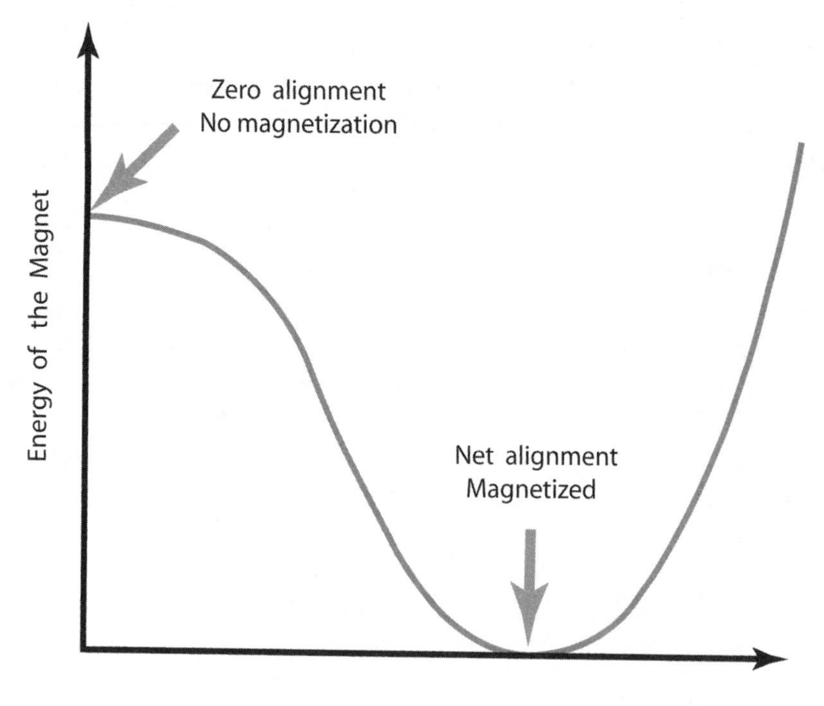

Figure 6.23. Magnetic Potential. The energy of an iron magnet as function of its magnetization shows that the minimum energy occurs for a nonzero value of the magnetization. This is why magnets form a stable state with a nonzero magnetic field.

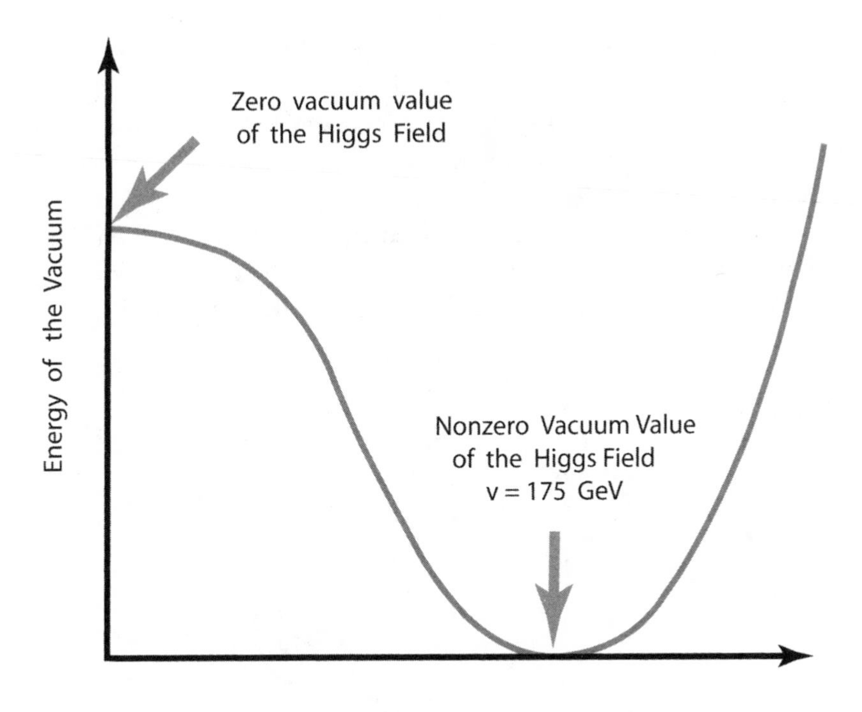

Figure 6.24. The Higgs Potential. We can adapt the principle of an iron magnet to the vacuum energy of the Higgs boson. The energy of the vacuum has a minimum for a nonzero value of the Higgs field. This causes the vacuum to develop a nonzero Higgs field everywhere throughout space. The parameters are tuned to produce a Higgs field strength in the vacuum of 175 GeV, the value that is inferred from Fermi's theory of the weak interaction and the muon lifetime.

We use this idea, theoretically, to make our Higgs field fill the vacuum. A vacuum in which the Higgs field is zero is simply "engineered" to be a state that has a higher energy than one in which the Higgs field has a nonzero value. Theorists know immediately how to do this: one simply relabels figure 6.23 for an iron magnetic, replacing "energy of the magnet" by "energy of the vacuum," and "magnetization" by "Higgs field." In this way we get the (now-famous) "Higgs potential" shown in figure 6.24. The preferred value of the Higgs field is just the location on the x-axis of the minimum energy point of the potential. This is the value the Higgs field will have throughout all of space. We can determine what it must be from Fermi's original theory. It is an energy and has a value of about v = 175 GeV. Voilà!

Katherine: "But once we have filled the vacuum with an enormous Higgs field, why can't we just go out in our backyard and pluck a Higgs particle from the vacuum? Why do we need the LHC at CERN?"

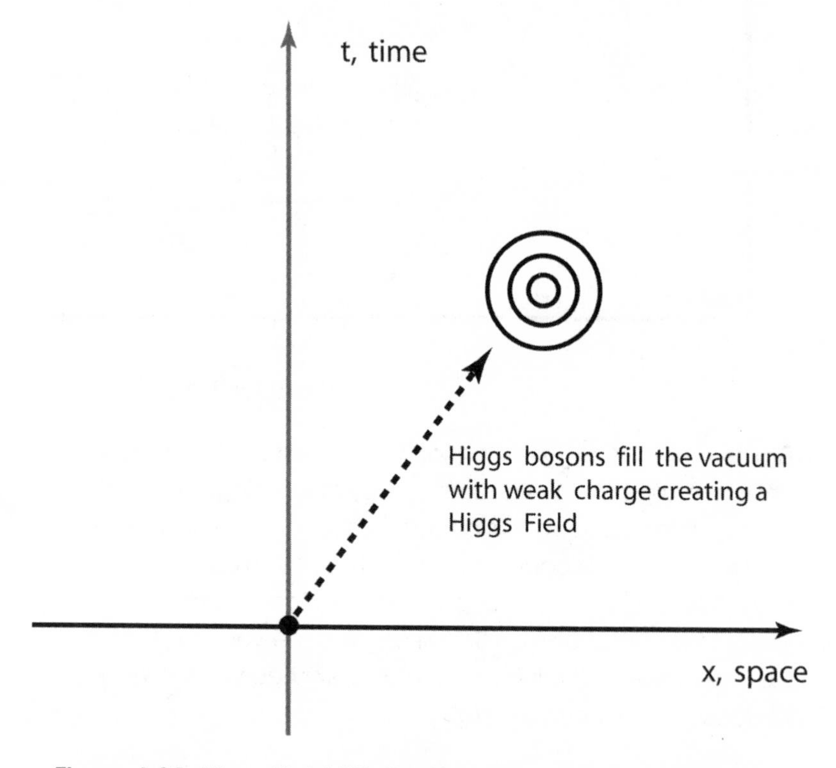

Figure 6.25. Higgs Field Fills the Vacuum. Higgs bosons form a field in the vacuum, represented by the circles, much like photons can form an electric or magnetic field. The Higgs field contains an indefinite number of bosons and an indefinite weak charge. A Higgs boson particle that undergoes a quantum fluctuation into a state of zero energy and zero momentum can "disappear" into the field (or "appear" from the field). The vacuum becomes a reservoir of weak charge.

Very good question, but it has a simple answer. The photon, which makes large electric and magnetic fields, has zero mass, so it isn't too hard to pluck a photon out of a large electromagnetic field. For example, we can have a source of light, like a laser beam, that is full of photons and that looks very coherent, as though there are no particles there. But we can also spread the beam out and make it very dim and put a "photon counter" or photo-cell hooked up to a computer, and we then see "tick . . . tick, tick . . . tick . . ." as individual photons are counted. We've thus plucked the photon particles out of the laser beam. These are very low-energy particles that can easily be detected by a sensitive detector. In fact, that's exactly what the silver halide crystals in an old photographic emulsion do: they react to individual photons as particles and when developed, give us a pretty picture, like Ansel Adam's view of the Grand Tetons.

And, indeed, the Higgs field that permeates the universe implies that Higgs bosons, as particles, are lurking inside the vacuum. However, the Higgs boson particles that collectively make up the Higgs field are very heavy particles. It takes a big sledgehammer to knock one out of the vacuum, and that's exactly what the LHC is doing.

But there's something else really interesting about the Higgs field—its existence means that the vacuum is full of weak charge. Recall that the Higgs boson must have a weak charge of –1 because it couples to L (–1 weak charge) and R (0 weak charge). The Higgs field throughout space means that the vacuum has become an enormous reservoir full of weak charge. We can borrow weak charge from the vacuum and, in so doing, turn a lowly R muon into an L muon. And an L muon can dump its weak charge into the vacuum and become an R muon. Eureka! We now see how the muon gets its mass by flipping from L-R-L-R. The flip involves a certain "coupling strength" of the muon to the Higgs boson, called g_μ. The mass of the muon is then determined: it is simply $m_\mu = g_\mu$ (times) 175 GeV.

The problem of giving the masses to the elementary particles is solved! The vacuum is an enormous reservoir of weak charge. R particles that have no weak charge absorb the weak charge from the vacuum to become L particles. L particles that carry weak charge dump their charge into the vacuum to become R particles. Since the great reservoir of weak charge, the Higgs field, fills all space for all time, the masses of the elementary particles are generated throughout all space and time. The gauge symmetry, i.e., the conservation of weak charge, is still in effect, operating at the microscopic level.

Muon acquires mass from interaction with Higgs Field

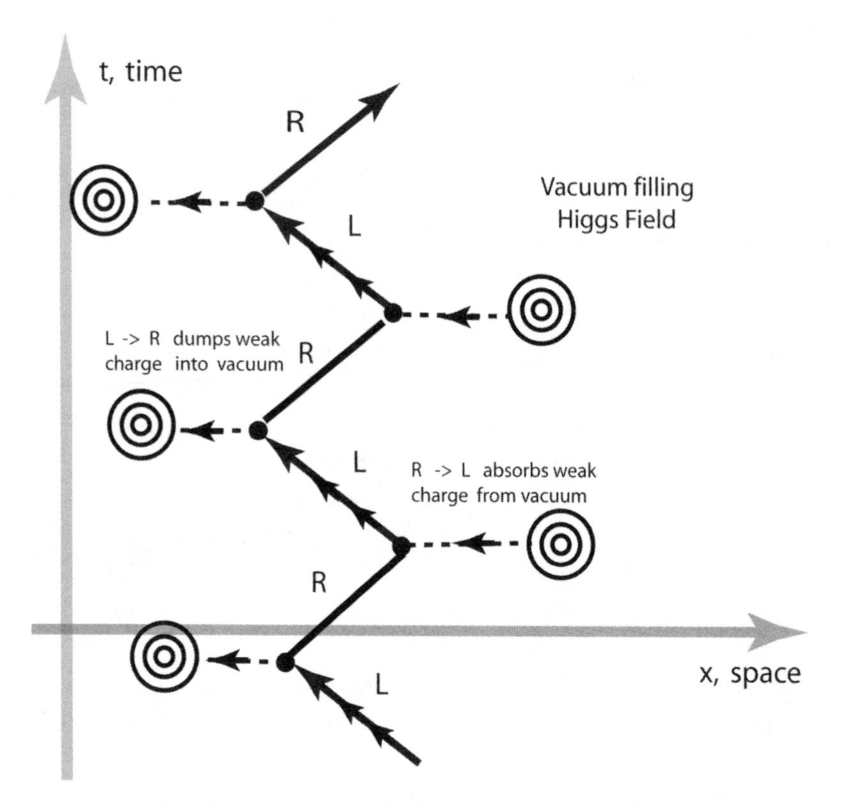

Figure 6.26. Muon Mass from the Higgs Field. The Higgs field, together with the coupling of the Higgs bosons to all particles, as in figure 6.22, gives rise to mass. The L muon can convert to an R muon and radiate a zero-energy Higgs boson that disappears into the field and carries away weak charge. In the figure the multiple arrows show the flow of weak charge out of the vacuum as R converts to L, then back into the vacuum as L converts to R. This also happens for electrons, the other leptons, and the quarks.

Alas, aside from solving the conceptual problem of making mass while preserving gauge symmetry, the formula $m_\mu = g_\mu$ (times) v, where v = 175

GeV/c^2 tells us very little. The problem is that the Standard Model doesn't *predict* the value of g_μ for us or the coupling of any other fermions to the Higgs boson,for that matter. All we can do is to go out and measure the mass of the muon, m_μ, and from that we can determine $g_\mu = m_\mu/v$. The measured mass of the muon is about 0.1 GeV/c^2, giving $g_\mu = m_\mu/v = 0.1/175 = 0.00057$. Likewise, the electron has a measured mass of about 0.0005 GeV/c^2, and therefore the electron couples to the Higgs boson with coupling strength $g_e = 0.0005/175 = 0.0000028$. The top quark has a measured mass of about 172 GeV/c^2, and therefore has a coupling to the Higgs that is nearly 1, i.e., $g_t = 172/175 = 0.98$. Here we've simply swapped the unpredicted masses of these particles for the unpredicted coupling strengths to the Higgs boson, g_μ, g_e, g_t, etc. But the Standard Model, through the Higgs boson, also generates the heavy masses of the W^+, W^-, and Z^0 bosons. These particles' masses were predicted by the theory, and the agreement between experiment and theory is spectacular. We leave the job of constructing a more complete theory that predicts the origin and mathematical values of the Higgs coupling constants to a future, younger generation.

We've arrived at the Rainbow Trout Lake, and we've just caught a big one. All particle masses, the L-R-L-R march through space-time, are due to the oscillation of L into R, and each time L turns to R it dumps a unit of weak charge into the vacuum via the Higgs field. And each time an R turns into an L it absorbs weak charge back out of the vacuum. *That is the origin of mass.* So, yes, Katherine, indeed you can "pluck a Higgs out of the vacuum," but to make and study the Higgs boson particle directly requires the sledgehammer of the CERN LHC. Our entire world is sculpted by the grand Higgs field that surrounds us. It's a little spooky.

LET'S TAKE A BREAK

Breathe in the cool and pure mountain air. We've fast-forwarded through the entire twentieth-century physics to get here. We deserve a break and a few moments to soak in the serene beauty of the mountain peaks and our lovely lake. We've just filled our basket with rainbow trout, and it'll soon be time to start back down the mountain. In fact, we are at the pinnacle. We now understand what the Higgs boson is and how its field fills the entire universe and how particles undergoing their L-R-L-R march through space-time are absorbing and reemitting weak charge, to and fro, into the vacuum itself. It's happening before us and all around us as we linger. All of the particles that make up our mountain, our lake, our trout, and us, are doing the L-R-L-R march and interacting with the grand Higgs-filled vacuum, dumping and absorbing weak charge as they go. And, on July 4, 2012, the experiments at the CERN LHC, finally confirmed to the entire world that it is true: The Higgs boson, the "particle comprising the Higgs field," the particles that collectively make up this vast hot tub of a vacuum in the universe in which we live, has finally been seen in the laboratory.

CHAPTER 7
MICROSCOPES TO PARTICLE ACCELERATORS

Our neighbors often ask, "So what do you folks do over there at Fermilab?" We would often tell them that "Fermilab has *the world's most powerful microscope*." This was true until November 20, 2009.

> "The LHC is back," the European Organization for Nuclear Research announced triumphantly Friday, as the world's largest particle accelerator resumed operation more than a year after an electrical failure shut it down.
>
> Restarting the Large Hadron Collider—the $10 billion research tool's full name—has been "a herculean effort," CERN's director for accelerators, Steve Myers, said in a statement announcing the success. Experiments at the LHC may help answer fundamental questions . . . which deal with [the particles of] matter far too small to see.[1]

The LHC was back after the challenging rebuilding process following its cataclysmic magnet explosion on September 19, 2008 (see chapter 1, under the heading "Oh, $%&#!"). At that time, the world's most powerful particle accelerator became the fully operational Large Hadron Collider (LHC) at CERN in Geneva, Switzerland. At that moment, a little-noted passage in history of profound significance had occurred: Europe became home to the world's most powerful microscope after nearly a century of US preeminence. The Fermilab Tevatron was switched off permanently on September 30, 2011. However, just as there are many kinds of microscopes, there are also many kinds of particle accelerators. Even today, Fermilab still operates many of the original onsite accelerators since the Tevatron shut down.

We think that the "the world's most powerful microscope" reply to our neighbors' question about the Tevatron then and the LHC now is the simplest and best "sound byte" one can give to the nonexpert—someone who may be curious about particle physics or about whatever mysterious things we do with these large devices. There were never any bombs built at Fermilab or CERN, and no UFOs are buried deep underground at the sites. It is all science, with the *biggest microscopes in the world.* The science of particle physics deploys accelerators, aka powerful microscopes, to study the smallest objects in nature—plain and simple. Our neighbors would usually remark at this point, "Oh! So that's what you do over there—how interesting," and then they might say something like "Hmm . . . I always thought it was something else."

Particle accelerators are, plainly and simply and precisely, the world's most powerful microscopes. To appreciate what these behemoths called particle accelerators are, let's take a look in detail at microscopes—let's put the microscope under the microscope.

MICROSCOPES

Ancients were aware of the phenomenon of lenses, but not much was done with them. The first serious practical application of lenses came with the invention of reading "spectacles" in the late thirteenth century, and primitive handheld "magnifiers" that were just tubes with a single lens at one end that could focus on a small object. An insect could be magnified by a few times, so these were often called "flea glasses."[2]

In 1590, two spectacle makers working in Holland (where they were probably counterfeiting coins), Hans Janssen and, particularly, his son Zacharias, discovered that with two lenses—one lens placed at each end of a tube—small objects could be made to appear greatly magnified.[3] This was the first *compound microscope* (several lenses), and it enabled magnification of small objects by about ten times. These early experiments also led to the telescope, which was perfected at about the same time, as was the science of optics, by Galileo. There is considerable uncertainty about the dates and attribution of these early developments of the microscope, involving a larger number of players (and no doubt considerable exaggeration and defamation), so we'll leave that to the historians. Certainly the development of the microscope was intertwined with that of the telescope. The telescope had such immediate importance to seafaring navigation that the microscope seems to have emerged as a secondary spin-off, much like the World Wide Web was a spin-off of particle physics.[4]

The celebrated "father of microscopy" and the "first microbiologist" was Anton Van Leeuwenhoek of Holland (1632–1723). Van Leeuwenhoek had no formal education, but he had considerable ingenuity and practical "street" skills. He apprenticed in a dry goods store where magnifying glasses were used to examine and count the number of threads in a fabric, and he later became a fabric merchant. Van Leeuwenhoek realized that high-quality lenses, with very short focal lengths, were needed to make better microscopes. This required extreme "double convex lenses," that is, almost perfectly spherical little balls of glass. These type of lenses are much more of a challenge to make than are the larger, less curved lenses required for comparable magnifications for telescopes. These extreme lenses, crystal clear and perfectly spherical, also demanded a higher quality glass, instead of the greenish "coke-bottle" glass of the day. Van Leeuwenhoek began to make lenses of pure crystalline quartz, "painstakingly teaching himself and developing arduous new methods to grind and polish" these tiny, near-perfect spherical lenses. In actual fact, Van Leeuwenhoek had evidently discovered some cleverly simple methods of manipulating glass to achieve these lenses. He may have deliberately given the false impression that it was only through tedious and skillful grinding methods that he could achieve these results to stave off competitors:

> Van Leeuwenhoek's interest in microscopes and a familiarity with glass processing led to one of the most significant technical insights in the history of science. By placing the middle of a small rod of glass in a hot flame, Van Leeuwenhoek could pull the melting section apart to create two long whiskers of glass. By reinserting the end of one whisker into the flame, he could create a very small glass sphere. These tiny spheres became the lenses of his microscopes, with the smallest spheres providing the highest magnifications. The shrewd Van Leeuwenhoek realized that if this simple method for making lenses was revealed, his role in microscopy might be minimized. He therefore allowed others to believe that he was laboriously spending most of his nights and free time grinding optically

> perfect tiny lenses to use in microscopes and he made about 200 high quality microscopes with different magnification powers.[5]

Ultimately, he could achieve a whopping and unprecedented magnification up to 270 times. The modern microscope was born, and with it the science of microbiology. With this powerful new and revolutionary scientific instrument Van Leeuwenhoek's discoveries were bountiful. He was the first human to observe and describe bacteria (obtained from scrapings off of his teeth), yeast, and other tiny plants and microorganisms. He was first to discover the protozoan animal life in a drop of water from a pond and the circulation of blood cells in small capillaries. His many discoveries were published in over a hundred letters to the Royal Society of England and the French Academy. "My work, which I've done for a long time, was not pursued in order to gain the praise I now enjoy, but chiefly from a craving after knowledge, which I notice resides in me more than in most other men. And therewithal, whenever I found out anything remarkable, I have thought it my duty to put down my discovery on paper, so that all ingenious people might be informed thereof."[6]

Van Leeuwenhoek's scientific contemporary was the great English scientist Robert Hooke (a contender with Isaac Newton for the theory of gravitation).[7] Hooke had likewise improved on early compound microscopes around 1660 and had made a number of the key early discoveries. In his book "Micrographia" (1665), Hooke coined the word "cell" to describe the basic building blocks of the internal structure of the plant tissues that he and Van Leeuwenhoek were able to observe under their microscopes. Hooke benefited greatly from the work, and was a champion, of Van Leeuwenhoek.

Van Leeuwenhoek had discovered the single-celled animals, or protozoans. How remarkable to think of it, this grand moment of discovery of the biological cell and the tiniest organisms of life, the constituents of all higher organisms, much like the atom is constituent of all materials! Hooke promoted the scientific reputation and career of Van Leeuwenhoek and sought to have him installed in the Royal Society. Yet, in 1676, trouble developed between Van Leeuwenhoek and the Royal Society, trouble that paralleled Galileo's earlier difficulties with the Catholic Church over the observation of Jupiter's moons through a telescope. The validity of Van Leeuwenhoek's discovery in a drop of water of protozoans, the smallest single-celled animals, was vehemently challenged, the issue revolving around evolution vs. religious preconceptions about life. Robert Hooke confirmed the discoveries, and only after considerable scrutiny and involvement of the English clergy was Van Leeuwenhoek ultimately vindicated in 1680. Observation and clear reason held the day, and Anton Van Leeuwenhoek became a Fellow of the Royal Society.

TECHNOLOGICAL CHALLENGE

The problem of making and using a nearly spherical lens is one of high technology, the highest of the late seventeenth century. A lens is always imperfect and creates a distorted image. One such distortion is the appearance of various fringes of color around boundaries in the image, called "chromatic aberration." This occurs, not because of imperfections in the fabrication, but because different wavelengths of light are refracted by different amounts when they enter or exit a piece of glass.[8] The resulting details of an image appear splayed out like a rainbow, yielding the color-fringed image in both telescopes and microscopes. This is essentially the effect by which glass can serve as a prism for resolving white light into its constituent colors.

By the mid 1700s a clever solution was developed to this chromatic aberration problem in telescopes, the so-called "achromatic compound lens." This is an arrangement of lenses that compensates the chromatic aberration and could be readily applied to the much less curved lenses in telescopes. But it remained a problem for the highly spherical microscope lens system. It was not until 1830 that Joseph Jackson Lister[9] (father of the famous surgeon Sir Joseph Lister, who was pioneer of antiseptic surgery) was able to make compound lenses to remedy the problem of chromatic aberration in microscopes.

Present-day optical microscopes are refinements of these early devices and give magnifications up to several thousand times. The perfection of high-quality-lens-based optical systems involves confronting and overcoming a number of various kinds of "aberrations." Most of these are dealt with by exploiting the idea of compound lenses, where two or more lenses are placed in-line and mutually compensate each other's aberrations (this is analogous to how a married couple can be better behaved than each of the individuals separately!).[10]

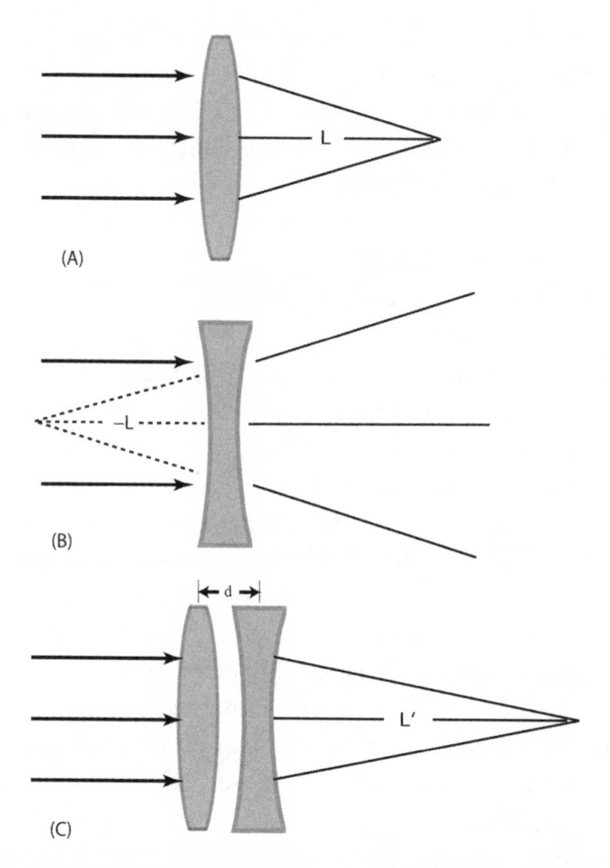

FIGURE 7.27. Lenses. (A) A double convex lens has focal length L, the distance from the lens at which parallel incoming rays are focused; (B) a double concave lens causes parallel rays to diverge from a point and can be assigned a negative focal length –L; (C) combining together (A) and (B), with a small space between the lenses, d, yields a compound lens with a net focal length of $L' = L^2/d$, so the compound lens, which is known as "FODO" (focus-space-defocus-space) is "net focusing." The compound lens can reduce aberrations of the two individual component lenses. This applies to magnetic lenses and forms the basic principle used in an "alternating gradient symnchrotron," where repeating quadrupole magnets form an ". . . FODOFODO . . ." continuous focusing system, also called a "FODO lattice."

The underlying idea of compound lenses involves a remarkable concept that turns out to be of foundational importance to the particle accelerators of the twentieth century. One can construct "convex" lenses that focus light to a point, modulo aberrations, and "concave" defocusing lenses that cause the beam or light to splay out. If a convex lens and a concave lens each having the same focal length are placed in-line close together but with a small air space between them, *the compound system is "net focusing."* That is, the compound lens, consisting of a focusing-defocusing pair, will have a much longer focal length than either of the two lenses have separately. The net effect is that light in the compound system will always focus, or converge, to a point, even though the convex lens is defocusing (see figure 7.27). The aberrations due to imperfections of the two separate lenses may be about the same, but they will mostly cancel out in the compound lens system. This results in tolerably smaller aberrations, much smaller than for a single lens with the long focal length. As we'll see, this net focusing effect of a focusing-defocusing compound lens is the basis for the existence of large particle accelerators, called "synchrotrons," such as the CERN LHC.[11]

HOW DOES IT WORK?

Let's consider how a microscope system works in very general and simple terms. We start with a "beam of particles." In the case of an optical microscope, this is a source of visible light. The particles are the *photons* that make up light (photons are quantum particles, so they behave both as particles and waves, but we'll put that aside for the moment). The source of the light, be it the sun or a candle or a lightbulb, is our "particle accelerator"—it has produced the particles, each of which carries energy as it moves toward us. We'll now put that beam to use in our microscope.

The microscope has a "target," typically a glass slide upon which you place the thing you want to see, such as a drop of pond water that contains the protozoans you wish to observe or perhaps a section of your tonsils (after they were removed). The incoming photons of our particle beam collide with the target. In these collisions the beam particles are scattered in all directions.

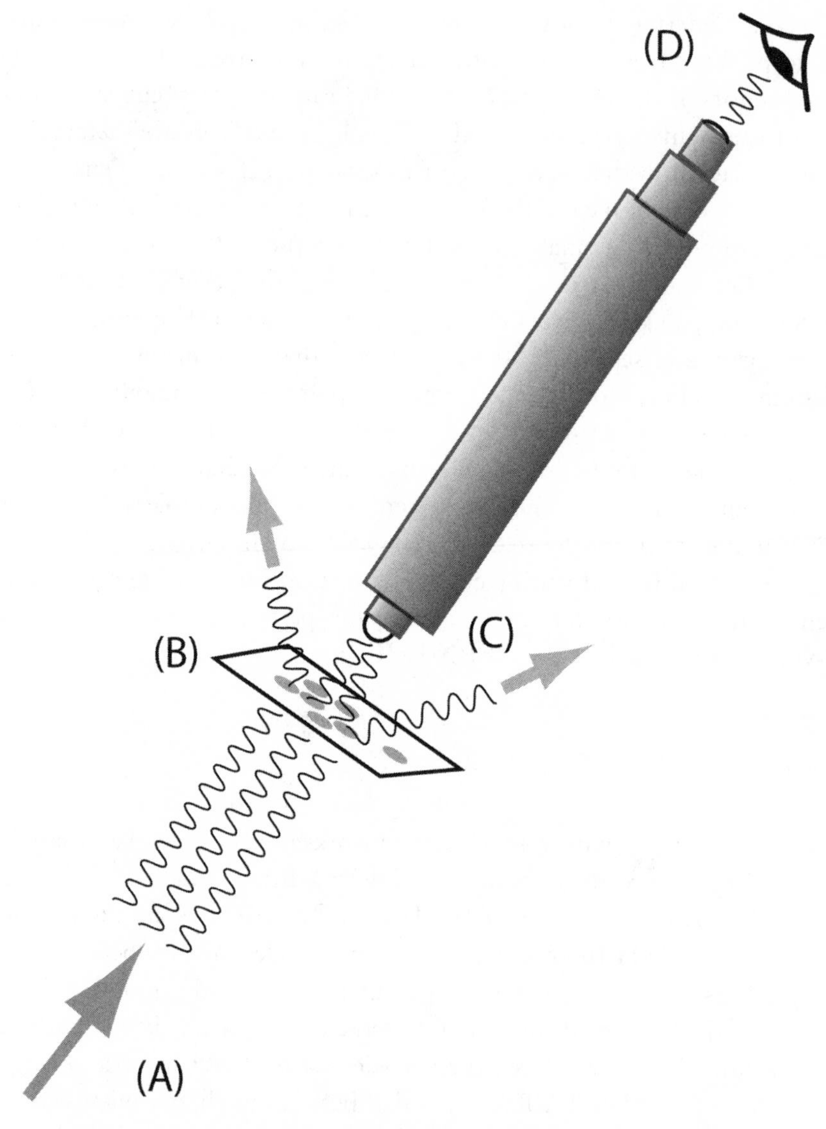

FIGURE 7.28. Microscope Schematic. Schematic of a microscope (or of a particle accelerator experiment). The system has (A) an incoming beam of particles, such as photons; (B) a target off of which the particles scatter; (C) a detector that collects and processes the scattered particles and presents the data to (D) the human observer.

Our microscope has a lens system. This is made of Van Leeuwenhoek's perfectly spherical crystal lenses in a configuration of Lister's chromatic aberration-free compound lens system. The lens system of the microscope collects some small amount of these scattered photons. It then focuses the collected photons to form an image, and the image is then presented to the eyeball of the observer.

So, the recipe for a microscope is

(1) Beam
(2) Target
(3) Detector = Lens system + eyeball
(4) Brain = Computer

OK, that's great! So now, let's just get a bigger and better Van Leeuwenhoek microscope with Lister's compound lens system and a bright beam of photons (higher luminosity), and let's go look at quarks! Unfortunately, it doesn't work quite that way. There are fundamental limits to how small an object we can see with an optical microscope. So, why can't an ordinary bio-lab microscope see a quark?

Recall our discussion in chapter 2 where we talked about the size of a probe vs. the size of the target: In an optical microscope the light particles, photons, are the probes. Therefore, what limits the resolution of a microscope is the physical "size of the photons" in our beam. In general, if the particles in our particle beam are "bigger" than the object we want to see, we will not be able to form a focused image. This is a kind of golden rule of physics: To measure something small, you need a probe of the object that is smaller than the object itself.

But, you say, photons are small, aren't they? They are elementary particles, and you said they have no discernible structure. They are pinpoints. So why won't they work as a probe? This is where we encounter the wavelike nature of all quantum particles. In short, the quantum theory paradoxically says that all things that are particles are also waves. They are both and they are neither. If this seems like an almost untenable logical paradox, all we can say is "Welcome to the quantum world, which no one really understands but our students are trained to use." This is the so-called "particle-wave-duality," and quantum theory is very subtle, yet very emphatic about the meaning of this. Even Einstein tried to rewrite this principle, but failed.

In the case of light, as we have discussed in chapter 2, photons are indeed point-like particles, with no size whatsoever, pinpoint, but they are also waves. And, it is the size of the *wave motion* of these pinpoint particles, or the "wavelength" of the quantum wave, that determines the "size of a photon" as far as microscopes are concerned (see figure 2.1).[12] As we've seen, ordinary visible light has wavelengths in a range of around 0.00005 cm (5×10^{-5} cm). Therefore, optical microscopes cannot resolve small structures below a scale of about 0.0001 centimeters (10^{-4} cm), no matter how well fabricated they are.

As the demands upon the science of microscopy moved to observe shorter and shorter distance scales, and the finest of lenses and optical systems were developed, this brick wall of the wavelength of light was encountered. Optical microscopes won't work to see the detailed structure inside the cells of living organisms, or to see viruses, large macromolecules, the DNA molecule itself, and so forth. To do so, we need a particle beam made of something smaller than visible photons.

MICROSCOPES THAT DON'T USE LIGHT!

The weird particle-wave duality of the quantum theory also provides a solution to the problem of building a better microscope.

First, the quantum theory tells us that *all particles* are simultaneously waves. This means we can use any particle we want, and it need not be a photon. For example it can be an *electron*, one of the easiest of all particles to find—electrons are found orbiting the nucleus of every atom in the universe. Second, the quantum theory tells us exactly how much energy we have to endow an electron with to dial up any particular quantum wavelength that we want. Third, electrons have electric charge, and it's therefore easy to give them an energy kick with the right kind of device. With a little kick in energy, we can create any tiny quantum wavelength we want for electrons. And, finally, electrons, and other charged particles can be focused with "electromagnetic lenses." Electrons are definitely particles, with mass and electric charge, and they have spin like little gyroscopes. Electrons are an ideal probe particle for the beam of a microscope.

The particle-wave duality of the photon was well understood by the 1920s, but it was thought that this property belonged exclusively to photons. It was therefore a real shocker when people first realized that electrons are also both particles *and* waves. This idea was due to a young graduate student, Louis de Broglie, who was studying the new embryonic quantum theory of Planck, Bohr, Einstein, Heisenberg, and others at the Sorbonne in Paris in 1924. It was also known, thanks to Niels Bohr, that electrons had a certain wavelike behavior when they were trapped in atoms. But it was thought that this had to do with the particular orbital motion of electrons when they were bound to atoms and that it was not necessarily an intrinsic property of the electron itself.

De Broglie proposed, in his PhD thesis, that the electron, like the photon, is a quantum particle-wave under all circumstances. It should therefore be possible to observe the wavelike motion of untrapped or freely moving electrons as they coast along through space. One should be able to do an experiment that reveals the characteristic features of waves with electrons. These general features of waves are known as *diffraction* and *interference*, common wavelike behaviors seen in light, or even in water waves. De Broglie wrote down the relevant equations in his brief, three-page-long doctoral dissertation at the Sorbonne—the equations are really quite simple once you get the basic ideas of quantum theory in your head.

The distinguished old-guard faculty of the Sorbonne was astonished by the brevity and simplicity of de Broglie's idea, but they were also unable to comprehend it. They were ready to dismiss the doctoral thesis altogether and send poor de Broglie home. Alas, and fortunately, someone sent a copy of his thesis to Albert Einstein with a request for a second opinion. Einstein replied that the young man in question deserved a Nobel Prize more than a doctorate degree.

Indeed, the wave motion of freely moving electrons was shortly thereafter confirmed in 1927 in an experiment in the US at Bell Labs by Joseph Davisson and Lester Germer. Electrons were seen to undergo diffractive interference, like light waves, as they bounced off the surface of a crystalline metal. This was a stunning development. No one had ever previously questioned that electrons were anything but hard little particles that would only scatter and bounce like billiard balls. Electrons, like photons, were proven to be waves—actually, quantum particle-waves—the enigmatic way of quantum theory. And de Broglie was, in fact, awarded the Nobel Prize in

Physics in 1929. The pieces of the quantum puzzle were soon put together into a new reality—*all particles are waves at the same time!*[13]

The question before us now is "What does this do for microscopy?" For charged particles such as electrons (charge = –1), protons (charge = +1), and even muons (charge = –1), etc., their quantum wavelengths can be made arbitrarily small if we can get these particles to arbitrarily high energies. Each of these particles conveniently has a special "handle" on it, the electric charge, that allows us to grab a hold and accelerate it. *Any electrically charged particle placed in an electric field will be accelerated and will acquire more energy of motion (kinetic energy).* The particle draws this energy out of the electric field.

In principle there is no limit to how high an energy we can give to an electron, or a proton, or even a muon, though for any acceleration scheme we eventually hit severe practical problems. Much of the modern development of accelerators involves overcoming various technical challenges (though, as we saw with the Super Collider, we couldn't overcome the political or financial challenges). In summary, using accelerated charged particles instead of light, we should be able to make "microscopes" of virtually unlimited resolving power and magnification.

PARTICLE ACCELERATORS

An accelerator is a device that takes some particles of matter, essentially at rest, and through the process of *acceleration*, endows them with a high kinetic energy (energy of motion). Recall that we always need a beam of accelerated particles as the first stage of a microscope system. The accelerator provides the beam.

A slingshot is a primitive form of particle accelerator. It consists of a wishbone-shaped frame, usually a branch cut from a tree, to which a large rubber or elastic band (the sling) is attached. The user places a stone in the band and pulls it back, thus stretching it and increasing the *potential energy* of the rubber sling, with the stone in the sling. He then aims at the target and lets go. As the elastic band snaps back into its original form, the stone is accelerated. The potential energy in the stretched elastic band (the energy vested in stretching the band) is converted into the *kinetic energy* (energy of motion) of the stone. Beware: Slingshots are *very dangerous* and you kids are advised not to play with them. For that matter, a gun is an accelerator, and it is much more dangerous. A car, a high-speed train, an airplane, a rocket ship are all people accelerators.

The physics of a slingshot is no different than that of a powerful proton, electron, or muon accelerator. The stone is replaced by a charged particle, such as an electron, and the sling band is replaced by an *electric field*.

ELECTRIC FIELDS ACCELERATE CHARGED PARTICLES

"Fields" are a central part of physics. You can't see fields, but they are there. They are real. They have energy, and they may or may not exert influence upon you. If a field has an influence on a particle, then we say that a particle is *coupled to the field*. "Electric charge" is the "coupling" of an electron or proton or muon to an electric or magnetic field. The field can then influence the motion of the particle, perhaps accelerating or de-accelerating it (electric field), or deflecting and diverting its direction of motion (electric or magnetic field).

The idea of "fields" began with gravity. Isaac Newton realized that there is a force of gravity between any two massive objects in the universe. He hypothesized that this force was attractive and proportional to the product of the masses of the objects. The strength of the force fell off with the separation of the objects according to the "inverse square law." He further hypothesized exactly how strong this force is for any pair of massive particles by a simple formula. The simple formula is called Newton's universal gravitational force law. The term "universal" means that you can plug into the formula the masses of any two objects, and their distance of separation, and out of the formula pops the gravitational force between the two masses. With some remarkable mathematical analysis (Newton also invented the "Calculus" to analyze motion in a gravitational force), he discovered that this one simple and elegant formula for gravitational force precisely explained the motion of the moon in its orbit about the earth, the motion of all the planets in their orbits about the sun, as well as the rate at which apples (or anything else) fall from trees on the earth. This was a "grand unification" of our understanding of the force of gravity, as well as the laws of motion of all objects in nature by gravity.[14]

About a hundred years after Newton's discoveries, it was realized that there are also *electric* and *magnetic* forces. It was in 1785 that Charles-Augustin de Coulomb discovered that a particle can exert a force (much stronger than gravity) upon another particle placed some distance away. Coulomb discovered that this force occurred when the particles have an attribute that was called "electric charge." Most matter is electrically neutral, i.e., it has no detectable charge, so we don't immediately see the effect of this new force. However, it is possible to generate net electric charge on objects, and then the powerful new force becomes apparent. The strength of the force between charged objects "fell inversely as the square of the distance between them," just like Newton's universal law of gravitation a hundred years earlier.[15]

The concept of electric charge evolved shortly thereafter, involving major insights of Benjamin Franklin and others.[16] We all know how it goes: a positive charge will attract a negative charge, while positive (negative) will repel another positive (negative)—likes repel while opposites attract—"in love as in electrodynamics." Today we also know that all elementary particles carry electric charges that can be measured and found to take on certain mathematical values. The charges found in nature are simple integer multiples of a basic quantity, called the fundamental charge, and that we call "e." A neutron, for example, carries zero times the fundamental charge and is electrically neutral. An electron carries minus one times the fundamental charge, i.e., an electron has charge –e, and a proton carries plus one times this charge, i.e., a proton has charge +e.

The universe has, as far as we can discern, an equal balance of positively and negatively charged particles. Most all charges have assembled themselves within ordinary matter into atoms, in which negative electron charges for the most part cancel positive proton charges. The electrical neutrality of matter is testament to the very strong nature of electric force compared to gravity—if we strip electrons off (or add them to) atoms we get *ions*, which are effectively atoms with an excessive, or un-canceled, positive or negative charge. Stray electrons can eventually find their way to an ion and combine to make a neutral atom, and the world more or less re-neutralizes itself.

A great debate ensued among physicists and philosophers as to what mediates the force "at a distance" between charges. There are no strings or springs or anything else visibly connecting them. This discussion in the early nineteenth century led to the concept of the electric field. The idea is that any electric charge actually produces a field that surrounds it. This field falls off in strength, following Coulomb's inverse square law (like Newton's inverse square law for gravity). A distant electric charge, which is *coupled* to the field by virtue of having a charge, will experience the presence of the electric field in the sense that it will be attracted to or repelled by the field, depending upon the sign of its charge. The force experienced by an electric charge in an electric field is determined only by the value of the electric field where the electric charge is located. And the electric field itself is produced by other charges located somewhere else, such as in the copper plates that make up the particle accelerator itself. Very simple indeed (see discussion referenced in note 15).

This leads us to useful applications. One such application is to make a "uniform electric field." We do this by taking two parallel plates of copper and connecting a battery to them (see fig. 7.29). The result is that electrons are sucked out of one plate (the one hooked up to the positive terminal of the battery) and deposited into the other plate (the one hooked up to the negative plate). This movement of electric charges halts when a large electric field builds up between the plates. The electric field is opposing the battery, pulling very strongly on the electrons, trying to yank them out of the copper plate that has a surplus of electrons, and move them back to the other copper plate that has a deficiency of electrons, i.e., to "re-neutralize" the system.[17]

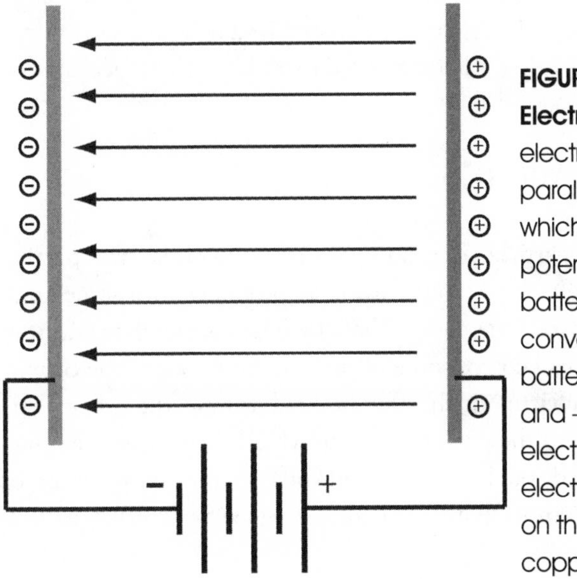

FIGURE 7.29. Uniform Electric Field. A uniform electric field between parallel plates of copper to which an electromagnetic potential is applied by a battery (note the use of the conventional symbol for a battery with indicated + and – terminals). A surplus of electrons, –'s (deficiency of electrons, +'s) accumulates on the negative (positive) copper plate.

However, cold copper metal holds on dearly to electrons. The electric fields among the copper atoms themselves cause these electrons to stay in the metal, and it takes a very large external electric field, or some thermal energy, i.e., heat, to overcome it. If we heat the metal, we are essentially injecting a lot of energy into it, and through random collisions we will kick electrons out of the copper plate. An electric current will then begin to flow between the plates. Electric current is just the motion of many charges together in some direction, and the electric field will become diminished as the electrons balance out among the two plates. Incidentally, this is best done in a vacuum to avoid the complications of air. With air present, the current refuses to flow (air is a *resistor* to current flow) until, eventually, the air molecules break down in some random channel in the air (the molecules ionize, i.e., they lose their electrons). The air then becomes a *conductor* of electricity, and the current will then quickly flow between the copper plates through this conducting channel of ionized air. That's called a "bolt of lightning."

If we set up our parallel plates in a vacuum, enormous uniform electric fields can be generated between the plates. Still, if the electric field becomes too great (perhaps about one hundred million volts per meter) the electrons will be ripped out of the copper (this is called "breakdown"), so we'll assume that we keep our electric fields below this critical level. Then, any lowly electron that is released into the vacuum at the negative plate will accelerate toward the positive plate and gain kinetic energy, like a rock in a slingshot.

We now have the basic ingredients needed to build a particle accelerator.

THE ELECTRON MICROSCOPE

The simplest charged particle accelerator, indeed, involves two parallel copper plates in a vacuum. We establish, using a power source of some kind, a large electric field between the plates. Then, by heating the plate with the surplus of electrons, we can coerce some of them to jump out of the copper. These will accelerate, by virtue of the electric field, toward the other plate. If we make a small hole in the positive plate to which they are accelerating, some electrons will pass through the hole, and voilà, we now have a beam of energetic accelerated electrons flying through space (see fig. 7.30).

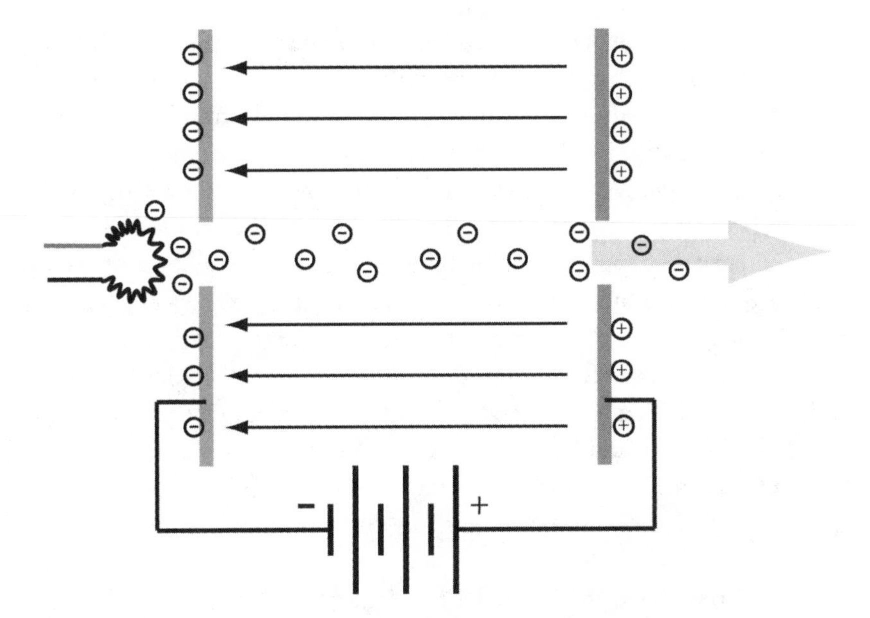

FIGURE 7.30. Acceleration of Charged Particles. A simple particle accelerator. Electrons are thermally ejected from a hot filament (zigzag loop on left) and pass through a hole in the negative plate. They are then accelerated in the electric field between the parallel plates toward the positive plate. They exit through a small hole, producing a particle beam with energy equal to the voltage difference applied to the plates by the battery (e.g., each electron acquires 12 electron volts of energy if we use a 12-volt battery). The system must be placed in a vacuum to allow the electrons to accelerate, unobstructed by collisions with air molecules.

This is a forerunner of most vacuum tubes that were used in early days of electronics. This ultimately led to the "picture tube" or "cathode-ray tube." Indeed, the cathode-ray tube is a simple particle accelerator. We used to say that everyone had a particle accelerator in their living room—the TV picture tube—but today cathode-ray picture tubes have gone the way of the horse-drawn carriage and the steam locomotive (replaced by thin digital screens that use light-emitting diodes or plasmas). However, there remain many sophisticated variations on this idea.

The principle involved to produce the energetic beam of electrons would work for any charged particles: protons, ions of atoms (atoms missing an electron), muons (a heavy sister particle to the electron), etc. It is the most basic principle of electromagnetism: a negatively charged particle is attracted to a positively charged particle and is repelled by a negatively charged particle—the potential energy, or voltage difference, between the plates (energy that comes from the battery) is the analogy of the stretched band of the slingshot. As the particle is accelerated, it takes energy away from the plates and acquires energy of its own, "kinetic energy," exactly like the rock in the slingshot acquires kinetic energy from the potential energy of the stretched sling. The cathode-ray picture tube employs about 20,000 volts power source, similar to that used in modern electron microscopes. The electrons are accelerated toward the screen of the picture tube and have been given 20,000 eV of energy each by the picture tube's "electron gun" (aka "particle accelerator").

So we now have the desired first element of a microscope: a small probe particle, which is an accelerated electron, and its quantum wavelength is many thousands of times smaller than a visible photon. We can therefore build an electron microscope by using a beam of accelerated electrons. The quantum wavelength of electrons can be shrunk to many times smaller than a human cell, even down to the size of an atom.

Recall our basic microscope design principles: (1) Beam, (2) Target, (3) Detector (eyeball), (4) Computer (brain). Here we use the electron beam to illuminate a target, which contains a specimen that we want to examine. The electrons collide with the specimen and are scattered in all directions. We then use a clever array of electric fields (and even magnetic fields, discussed below) to make a lens that focuses the scattered electrons—the detector. "Electromagnetic lenses" play the same role as the glass lenses of Van Leeuwenhoek in an optical microscope. The lenses form a magnified image from the scattered electrons that have collided with the specimen. Since our eyeballs aren't designed to see electrons hitting our retina (not a good thing to do), we project the electrons onto a phosphorous screen, much like a TV picture tube (nowadays the electrons are detected in a silicon array and the data is processed by a computer). In this way we get a beautiful magnified image of our specimen.

An electron microscope has much greater magnifying power than an optical microscope because electrons have quantum wavelengths that can easily be about 100,000 times shorter than visible light (photons). Electron microscopes can achieve magnifications of up to about 10,000,000×, whereas ordinary optical microscopes are limited by the wavelength of light to useful magnifications below 3,000×. What a spectacular new world opens up to the electron microscope![18]

Electron microscopes are used to observe a virtual infinity of biological and inorganic specimens. These include microorganisms, bacteria, viruses, plasmids, cells, large molecules, biopsy samples, metals, crystals, agricultural vegetable and soil samples, fragments from archaeological or geological sites, microelectronic circuits and chip fabrication as used in computers, forensic stuff from crime scenes—you name it. Industrially, the electron microscope is used for quality control and failure analysis—checking out the metal and rivets in the wings on aircraft and the support structures of buildings and bridges—studying how metals become fatigued in everything from gasoline engines to battery terminals. Modern electron microscopes usually don't use picture tube screen displays anymore and instead produce electron micrographs using specialized electron-detecting circuits (making pixels) and digital displays with computerized image-processing and image-capture capabilities. Fewer accelerated electrons with more computer image processing means the sample (target) life is longer, and even motion pictures become possible.

The social and economic impact of this, the simplest scientific particle accelerator known as the electron microscope, is profound and enormous. Since neither of the authors have PhDs in economics, we are reluctant to assign a dollar value to this, but we are sure it is many, many times what our nation is spending on basic research and the world's highest-energy particle accelerators today. This is a pure example of how investment in basic science has created and sustained our economy. What science has given to our economy and its standard of living through this simple device, the simplest of particle accelerators, the *electron microscope*, could probably have paid for many Super Colliders and most of the science funding in the world today.

CHAPTER 8

THE WORLD'S MOST POWERFUL PARTICLE ACCELERATORS

"I don't know what good this will be, but one day you may tax it." So said Michael Faraday to the Chancellor of the Exchequer, William Gladstone, who visited his lab to see where all that "wasteful spending" on physics was going. In Faraday's lab the money was going to develop wire and coils and batteries, or, in short, to invent "electricity." Faraday studied the property of electrons interacting with electric and magnetic fields, the particle physics of his day, and he laid the groundwork for the post–steam engine Industrial Revolution, the modern era, based upon electricity.[1]

Particle accelerators are a mere slice of the pie of the full range of applications of Faraday's science of electromagnetism in the general economy. We can only guess how much revenue particle accelerators contribute today to the US GDP. They are ubiquitous, found in most hospitals, laboratories, universities, and most high-tech industrial centers. They are a direct spin-off of basic research in the twentieth century of the study of atoms, nuclei, and elementary particles. We suspect that the federal revenues derived from the taxing of these facilities are far greater than those being spent on R&D of new and larger particle accelerators and their technologies. In fact, most of the federal money spent on particle physics is for the research, construction, and operation of accelerators and the physics experiments that use them.

We're going to survey the basic kinds of machines that are used in particle physics today, and that can be construed as the world's most powerful microscopes. Bear in mind our microscope analogy: the accelerator produces the beam that we shine on our target to scatter into the detector, which makes an image for our brain to comprehend. The beam particles must be smaller than the things we wish to observe, i.e., they must have smaller quantum wavelengths, and this demands high-energy particle accelerators to produce the beam. The collisions in "colliders" occur among two opposing beam particles themselves. The eyepieces of the microscope are large detectors, as large as a mansion, which are literally wrapped around the point of collision. Rather than one guy in a white coat staring through an eyepiece at a bacterium, high-energy particle experiments are a legion of physicists staring into computer screens and tweaking computer code. But it's all still basic microscopy.

LINACS

With the electron microscope we encountered our first "linear particle accelerator," often called a linac. A high-energy linac accelerates charged subatomic particles, such as electrons or ions (ionized atoms) in a straight line.[2]

The energy of a particle determines its quantum wavelength, which we want to make small. To shrink by half the wavelength of a particle traveling near the speed of light, we need to double its energy. To do so requires either a longer linac or a stronger accelerating field. But it is impossible to make arbitrarily strong electric fields. A really strong electric field will yank electrons out of the materials that produce the electric field, such as copper plates. This produces a flow of electric charge that neutralizes the electric field. Since our particle accelerators already use the largest electric fields we can possibly make, to double the beam energy of a linear accelerator we therefore need to expose the particle to the same electric field for twice the distance, that is, we need to double the physical length of a linear accelerator to double its energy.[3]

We measure electric fields in terms of *volts per meter*. An electric field always exists in the air on a warm, humid day, pointing upward to the sky, typically of about 100 volts per meter. You don't even notice this field, but when a big thundercloud rolls by, the field (e.g., particularly around metallic, pointed objects) can become many thousands of volts per meter. Electrons are then ripped out of materials. This is, in fact, the principle of a lightning rod—it is designed to allow electrons to drain out of a metal point because this tends to neutralize the surrounding electric field and thus avoids a lightning strike. If the electric field should fluctuate up to hundreds of thousands of volts per meter near an object, then enough air molecules can become ionized to create an electrically conducting channel—and then you get a spectacular lightning bolt.

In modern particle accelerators the accelerating electric fields are almost always generated in copper or alloy superconducting radio frequency (RF) "cavities."[4] These cavities are to electric fields essentially what a bell or a guitar string is to sound. We can fill the cavity with a rapidly oscillating electric field, which is like the sound of a ringing a bell or the plucking of a guitar string (the cavity-filling fields are generated from sophisticated devices, which are themselves mini-accelerators, called "klystrons"). The electric field then vibrates, or *resonates*, within the cavity. The walls of the cavity simply contain the electric field and offer essentially zero electrical resistance, which would otherwise damp out the field (like putting your finger on a vibrating bell or guitar string). The wavelength of the resonating electromagnetic field is then typically twice the size of the cavity, like the wavelength of the sound emanating from a vibrating guitar string.

If an electrically charged particle enters the filled RF cavity at just the right moment in the ringing cycle, it will experience an electric field that accelerates it through the cavity. We say that the charged particle is "in phase with the electric field." The particle is then pulled along by the field and will absorb kinetic energy from the electromagnetic field in the cavity ("out of phase," and the particle would be pushed back). Often people make the analogy to a surfer catching a wave. This analogy is picturesque, but there's a big difference. In our case the electron acquires more and more (kinetic) energy as it experiences the pulling of the electric field, while the surfer, once in motion, rides along at the peak of a water wave at an approximately constant kinetic energy.

The highest useful electric fields achieved in microwave cavities are about 30 million volts per meter. At still higher fields, the physical limits are reached, and copper becomes damaged by the ripping out of electrons, or the superconducting state of the material is destroyed. This presents a physical limit on electric fields that we can achieve in the lab or use in a device.

To build a high-energy linac we can use the clever idea attributed to Leó Szilárd and patented in 1928 by Rolf Widerøe, who built the first linac. We simply string a large number of RF cavities together in a line.[5] We arrange for these to resonate so that a charged particle passing through the sequence of cavities always "feels" an electric field that pulls it along the way. We inject particles, usually from a simpler smaller accelerator that serves as an injection device, into the first cavity. After getting a kick of energy from the first cavity, the particle then enters the next one to receive another kick, and so on down the line. In this way, through a sequence of energy kicks, we can create a very high-energy beam of charged particles.

A linac can be configured to accelerate different kinds of particles, which may be electrons, protons, ions (heavy atoms with a net electric charge), or even unstable elementary particles such as muons. The only requirement is that the particles we are accelerating must have electric charge so they "feel" the electric fields. If we desire high-energy neutral (uncharged particles) particles, like neutrons or neutrinos or neutral π^0's, we can make them by colliding our high-energy charged particle beam into a block of material, like aluminum, tantalum, lead, or even uranium or a spray of liquid mercury. The collision of our primary beam particles with the atoms within these materials will produce these various other types of particles.

Linacs range in size from the cathode-ray tube in an old TV picture tube, to a meter-long electron microscope, to the 100-meter-long proton linac at Fermilab, way up to the 2-mile-long electron linac at the SLAC National Accelerator Laboratory in Menlo Park, California. Like their relative, the electron microscope, linacs also have many practical applications, such as generating X-rays and gamma rays and high-energy electrons for many material science studies and medical radiation therapy.

Today, the 2-mile-long accelerator at SLAC has been retired from particle physics, yet it continues to serve science as a Linac Coherent Light Source (LCLS). The LCLS uses part of the former linear accelerator and is the world's first "X-ray free-electron laser." It produces pulses of X-rays each only a billionth of a second long yet a billion times brighter than any other X-ray source. This can be used to take instantaneous pictures of atoms and molecules in motion, such as in a chemical reaction, perhaps in a living cell. This provides crucial information on fundamental processes of chemistry, biochemistry, and technology.[6]

Linacs at Fermilab and CERN accelerate protons (or the H^- ion) and are used as particle injectors for the higher-energy accelerators. A single RF cavity, about a meter long, can be a stand-alone source of about 30 MeV electrons that can be used for medical and material industrial applications.

Linacs, unlike other circular accelerators, are capable of an output of many, many particles, i.e., a very high-beam current or intensity, producing a nearly continuous stream of particles. The potentially high intensity makes the proton linac an ideal accelerator for studying rare processes—of elementary particles, nuclei, or atoms, where the energy per particle is not as important as the sheer number of accelerated particles.[7]

Fermilab has plans to build a linac that will be the world's most intense source of protons in the world, called "Project X." Each proton in Project X will have a mere 8 GeV of energy, (compared to 7 TeV at the LHC; recall that 1TeV = 1,000 GeV), but the machine will have many, many protons and will produce a beam of very high power. We'll have more to say about Project X in chapter 9.

SCRFS: THE NEXT BILLION-DOLLAR THING FOR THE ECONOMY?

After the Superconducting Super Collider (SSC) perished, many US particle physicists set their sails to try to convince the US government to build an enormous linear accelerator that would collide electrons and antielectrons head-on. Despite shockingly high cost estimates and a lukewarm reception by the Department of Energy to undertake such a project, the International Linear Collider (ILC) community has forged on with their dream, creating an official management structure called the "Global Design Effort."

In 2004, the ILC Global Design Effort made make a monumental decision: The ILC would be based upon the new technology of *superconducting radio frequency cavities* (SCRFs.) This decision was risky because these cavities had not yet been developed for reliability or to a large-scale manufacturing capability. While this concept had, *a priori*, a number of technological advantages, it was also fraught with the risk that it required a long-term and expensive R&D effort to learn how to make SCRF cavities. But this was a brave decision that has enormous spin-off potential for the betterment of humanity. The US Department of Energy (DOE) has now invested an order of a half billion dollars on the development of this new technology.

The R&D has been a spectacular success in leading to the development of reliable SCRF cavities. SCRFs are very compact and efficient particle accelerators in terms of the energy they consume. The "wall-power" efficiency, defined as the fraction of energy that ends up in the electron beam compared to the energy consumed from the power source, is many times greater than conventional non-superconducting cavities. To give a sense of required reliability of SCRFs in this R&D program, a 500 GeV ILC requires 18,000 nine-cell RF cavities.[8]

Whether or not there will ever be an ILC (and there is currently talk of a collaboration between the US and Japan to do so), the potential benefits to humanity of the development of SCRFs are enormous. Though the SCRF accelerates an electron (or proton) beam, the output beam can readily be converted to gamma (high-energy photons) or neutrons. Potential applications include many aspects of medical imaging and medical therapy.

For example, an urgent need exists for the onsite hospital production of the isotope technetium-99 (Tc^{99}). This is the most-often used imaging isotope in medical diagnosis. Tc^{99} has a half-life of a few hours and must usually be prepared at the hospital site from other radioactive sources, but this presents a challenging and somewhat unreliable supply chain.[9] SCRFs are a logical candidate to provide a solution to this problem, which would probably eventually pay back many times over a year the cost of the entire R&D program that led to SCRF development. Electron beams can also be used for the removal of noxious gases in flue gas (chimney smoke) such as SO_2 and NO_2, surface treatments of materials, novel tunneling techniques, etc. It's "blue sky" for the future of SCRFs, and we expect to see a multibillion-dollar industry in the not-too-distant future.

And once again, Mr. Gladstone, you will tax it.

MAGNETIC FIELDS MAKE PARTICLES MOVE IN CIRCLES

As we've seen, a charged particle can be accelerated in a line by an electric field. This acceleration imparts energy to the particle and shrinks its quantum wavelength to enable the world's most powerful microscopes, aka particle accelerators. However, like electric fields, magnetic fields also accelerate a charged particle, but *only if the particle is already moving* and *only perpendicular to the direction of motion.*[10] The net result of this perpendicular acceleration is that the charged particle in a magnetic field will move in a circle, but it will acquire no net increase in energy, i.e., it will only be "deflected" from the original direction of motion.[11] Therefore, a magnetic field cannot be used to impart energy to our beam particles and shrink its quantum wavelengths, but it can be used, indeed it is essential, for *steering and focusing charged beam particles.* Magnetic fields, by deflecting a particle's motion, allow us to create an *electromagnetic lens* and to make a focusing lens system for a beam of particles. This is precisely what is done in electron microscopes to focus an image.

However, there is another and very important thing we can do: by cleverly using magnetic fields to bend charged beam particles in a circle, we can build a much higher energy particle accelerator than a linac that is more compact in size and therefore usually less expensive. Such circular machines, called cyclotrons and synchrotrons, exploit circular orbits of charged particles moving in magnetic fields to cause the particles to pass many times through the same RF cavity and receive many kicks in energy. The magnetic field simply holds the particle in its circular orbit.

As the beam particle's energy increases, and if the magnetic field is held constant, the particle will stray into a larger and larger radius orbit. To hold the particles in the same orbit, the magnetic field must also increase with the beam energy. This is the principle of a particle accelerator called a *synchrotron.* As there are limits to the highest electric fields we can create, there are also limits to how large a magnetic field we can create. This translates into the requirement of very large-diameter circular orbits for very high-energy particles. For example, the Tevatron achieved a 1 TeV (one trillion electron volts) beam energy for protons and antiprotons, and was a circular ring about 1 mile in diameter; the LHC is designed to achieve a 7 TeV beam particle energy, uses slightly stronger magnets, but is about 5.3 miles in diameter.

ELECTRIC CURRENTS PRODUCE MAGNETIC FIELDS

As we've seen, electric fields are generated by electric charges and cause electric charges to accelerate. Likewise, magnetic fields are generated by electric currents, and they in turn cause electric currents to deflect in direction or move in a circle.

Moving electric charges are called electric currents. André-Marie Ampère discovered that magnetic fields are produced by electric currents, and that electric currents are affected by magnetic fields.[12] Electric charges that are not moving do not produce magnetic fields and are not affected by them. Electric currents in matter, such as in copper or aluminum wire, consist of the most loosely bound electrons moving through the material to produce the current. As far as the electric charge goes, this is an electrically neutral situation—all positive electric charges are sitting at rest, such as the atomic nuclei with most of the electrons remaining attached to the atoms. The looser electrons can be coaxed to move through the material by a battery and become an electric current. The stationary charges associated with the atoms do nothing but keep the material electrically neutral, so that the net electric charge of the material is zero even though it may be carrying a large electric current.

In a simple experiment, two wires with constant electrical currents flowing in them, when placed parallel and near to each other, will be seen to attract or repel each other.[13] If the currents are moving in opposite directions, the wires repel each other; if they are moving in the same direction, the wires attract each other. This is a direct observation of the connection of magnetism to electrical currents. The wires are electrically neutral (no net electric charges, hence no electric fields are present), but it is the electric currents that produce the magnetic forces between the wires. One wire carrying a constant electrical current will also cause a compass needle to deflect.

The ancients, like many people today who are not familiar with magnetism, viewed it as a mysterious, almost magical, property of certain materials, such as iron or "lodestone" (which contains iron). The Chinese were the first to note the existence of magnetism and apply it to build a compass.[14]

An electron by itself, unattached to an atom, produces its own tiny magnetic field due to its intrinsic spin. In most materials the electrons are paired, in opposite spins and opposite orbital motions, and the magnetic field of one electron is canceled by an opposite magnetic field produced by the other electron in the pair. The atoms in materials such as iron, cobalt, and nickel have unpaired electrons. As a result, though each atom of these elements acts like a very small magnet, if all the magnetic fields of all the atoms are aligned together, the effects can add up to make a large macroscopic magnetic field. In iron, the little individual atomic magnets within the material interact with one another in such a way that they have a tendency to line up in a common direction. This forms a "magnetic domain," in which clusters of millions of atoms align to produce a common magnetic field. Finally, with a little more coaxing, usually an external applied magnetic field, the domains themselves can be coaxed into alignment, and then you have a powerful magnetic field emanating from a bar of iron. The physics at the atomic level of all of this is quite subtle, complex, and an interesting subject to study—so, yes, in a sense magnetism is a mysterious and special property of certain materials, but it isn't magic.

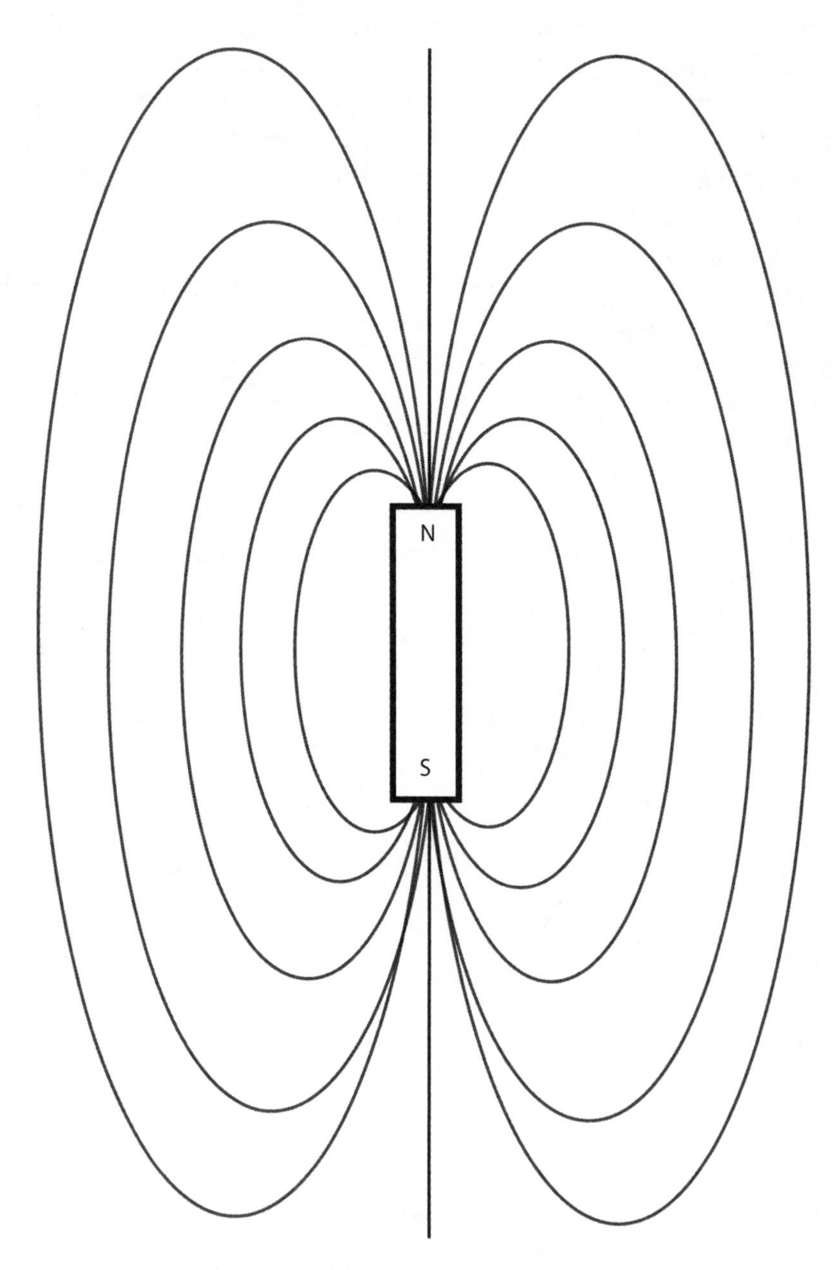

FIGURE 8.31. Bar Magnet. The magnetic field of a bar magnet.

The poles of a magnet are called north (N) and south (S). If a small iron bar magnet is hung from its middle by a string, it becomes a compass needle, and its N end will point northward, thus its S end points southward. The N end of a magnet will repel the N end of another magnet, S will repel S, but N and S attract each other. Hence, N and S are like positive and negative charges (we call N and S magnetic monopoles). However, we can never have an isolated N without, somewhere, having a compensating S; all magnets are therefore "dipoles," i.e., having two opposite poles, with equal but opposite N and S. This is a consequence of the magnetic field being set up by electrical currents, rather than having magnetic charges, or "magnetic monopoles," as their sources.[15] Either pole of a magnet will induce magnetization in a nearby magnetic material. Therefore, either pole can attract iron-containing objects, such as paper clips, because the magnet will induce magnetization in the paper clip. The paper clip becomes itself a temporary magnet, with its N pole facing an S pole, or vice versa.

If we arrange a flat white sheet of paper over a bar and sprinkle over the paper little iron filings, the filings will align with the magnetic field and allow us to visualize the magnetic field itself![16]

CYCLOTRONS

We'll only mention cyclotrons in passing, since they are rather passé in modern particle physics, and will instead refer the interested reader to the extant literature, e.g., search online for "cyclotrons" or see the *Wikipedia* entry.[17] The idea of a cyclotron is to accelerate charged particles but to hold them in circular spiral motion with a *constant magnetic field.* For example, we can inject particles into the center of a circular machine with a perpendicular magnetic field. We give the particles a little kick in energy, and they will move in a circle. Each time the particles complete one full turn, they are given another "kick" of energy from the same electric field, and then the cycle repeats. As the particle receives each kick in energy, it will tend to spiral outward into a circular orbit with a larger radius.

The cyclotron was invented in 1932 by Ernest Lawrence of the University of California, Berkeley, with much of the development in collaboration with his student, M. Stanley Livingston. The cyclotron was an improvement over the linac of the 1920s, when it was invented, being more compact and cost-effective due to the circular repetitive acceleration process.

For several decades, cyclotrons were the best source of high-energy beams for nuclear physics experiments; several cyclotrons are still in use for this type of research. Cyclotrons are still actively used in medical applications to treat cancer and to produce radioactive isotopes for medical imaging. Ion beams from cyclotrons can be used, as in proton therapy, to penetrate the body and kill tumors by radiation damage, while minimizing damage to healthy tissue along their path. Cyclotron beams can be used to bombard other atoms to produce short-lived positron-emitting isotopes suitable for PET imaging.

SYNCHROTRONS

The most advanced circular accelerator, and the one used most commonly in particle physics, is the synchrotron. The first electron synchrotron was constructed by Edwin McMillan in 1945, although the principle had already been published in a Soviet journal by Vladimir Veksler.[18] The first proton synchrotron was designed by Sir Marcus Oliphant and built in 1952. Protons, and even muons, can be accelerated to very high energies in large rings without appreciable synchrotron radiation loss. The Large Hadron Collider and former Tevatron were both proton synchrotrons (the Tevatron also circulated and accelerated antiprotons in the opposite direction in the machine).

In a synchrotron, particles are held in a fixed circular orbit. Each time they complete a cycle in the machine they receive a kick of energy from RF cavities. As the energy of the particles increases, the magnetic field is also slightly increased to maintain the same orbit. This allows the vacuum beam pipe that contains the motion of the particles to be a very large circular shape, rather than a large disk, as in the cyclotron. The smaller cross-section beam pipe allows for the magnetic fields to be localized within the pipe. These factors allow us to build very high-energy machines that are far less expensive than cyclotrons or linacs.

Synchrotrons "hit a wall" when they are used to accelerate electrons.

Owing to its small mass, an electron tends to radiate photons copiously when placed in a circular orbit at high energies. This is called "synchrotron radiation." To achieve high energies an electron synchrotron has to be quite large, or else most of the acceleration energy will be radiated away. The Large Electron–Positron Collider (LEP) at CERN that collided electrons with positrons was a synchrotron that occupied the large tunnel that today houses the LHC, with a diameter of about 5.3 miles. LEP pushed the limit of achievable synchrotron energy with a beam of electrons circulating in one direction and positrons in the other, each beam having about 100 GeV of energy per particle (hence 200 GeV total; LEP went a bit above this energy scale in its last days). Energies of about 45 GeV per particle create a total energy in the collision of 90 GeV, allowing direct production of the Z^0 boson. The synchrotron energy loss per orbit at LEP was about 0.2 percent. At the highest energies, about 100 GeV per beam, the electrons lost about 2 percent of their energy every time they orbited in the machine due to the synchrotron radiation. Synchrotron radiation itself is useful, however, in many applications, particularly in the study of chemical and biological reactions. Large circular electron machines are often designed to be sources of the high-energy photons from synchrotron radiation, or "synchrotron light sources."[19]

MAGNETIC LENSING

Recall that a lens can, ideally, focus all the photons that are moving parallel to the axis of the lens to a point. It turns out that we cannot make a magnetic field that focuses charged particles exactly like a lens. With a "quadrupole magnet," however, we can focus the electrons that are moving in one plane—let's say the "horizontal plane," but then we end up defocusing by the same amount in the perpendicular, or vertical plane. If we rotate the quadrupole about its axis by 90° we will get exactly the opposite: particles in the vertical plane are now focused, while those in the horizontal plane are defocused.[20]

However, here we exploit the fabulous trick that was used by microscope and telescope lens makers to correct for chromatic aberration. We can have one quadrupole that focuses (defocuses) in the horizontal (vertical) plane, followed by a second quadrupole, rotated by 90 degrees, that that focuses (defocuses) in the vertical (horizontal) plane. However, recall that when a focus lens (F) is followed by some space (O), which is then followed by defocus (D) and more space (O), the net effect is to focus. That is, a compound lens that is focus-space-defocus-space, or "FODO," *is net focusing* (see figure 7.27 caption)! We can therefore keep a tightly focused beam orbiting within our synchrotron by repeating this arrangement: FODOFODOFODO. . . . This technique is called *alternate gradient focusing*. This was the breakthrough that led to large synchrotrons, such as the first Alternating Gradient Synchrotron (AGS) at Brookhaven National Laboratory, then eventually to the Tevatron and LHC.[21]

Synchrotrons are unable to accelerate particles from rest, and they require a sequence of pre-acceleration stages. This can be done by a chain of other accelerators, like linacs or other smaller synchrotrons, and an initial kick from something simpler, like a high voltage.

Let's examine the most recent highest-energy colliders that have ever been constructed to see how this symphony of components perform.

THE WORLD'S GREAT COLLIDERS

TEVATRON

The Tevatron was a synchrotron particle accelerator at Fermilab.[22] It was the highest-energy particle collider in the world until it was superseded by the LHC at CERN. The Tevatron accelerated protons in one direction in the machine and antiprotons in the opposite direction, and brought them into collision at two points within the machine. At these locations were the detectors, the CDF and D-Zero, which played the role of the "eyepieces of the microscope." The Tevatron ring was 3.9 miles in circumference, and it accelerated the protons and antiprotons up to energies of 1 TeV, producing collisions up to 2 TeV in energy.

The Tevatron was completed in 1983, and significant upgrades were made continually during in 1983 through 2011. The Tevatron discovered of the top quark, made the most precise measurement of the W-boson mass (and initially a much improved measurement of the Z-boson mass, which was soon superseded by CERN's LEP accelerator), saw the first hints of CP-violation in "b"-quark physics, and made numerous other measurements concerning the strong interactions of quarks.

The first large accelerator at Fermilab was called the Main Ring, and construction on it began on October 3, 1969, with the groundbreaking led by the lab's first director, Robert R. Wilson (Fermilab was then known as the National Accelerator Laboratory). This would become the 3.8-mile-circumference tunnel that eventually housed the Tevatron. The Main Ring used conventional copper wire magnets and achieved a typical beam energy of 300 GeV by 1973, and a record beam energy of 500 GeV in 1976. To go to higher energies required more powerful magnets, and this required the technology of superconductivity. The Main Ring accelerator was shut down on August 15, 1977, and newly developed superconducting magnets and a new beam pipe were mounted on top of the old Main Ring magnets. The superconducting "Tevatron" produced its useful beam energy of 900 GeV in November 1986.

> On September 27, 1993 the cryogenic cooling system of the Tevatron was proclaimed an International Historic Landmark by the American Society of Mechanical Engineers. The system, which provides liquid helium to the Tevatron's superconducting magnets, was the largest industrial scale cryogenic system in existence upon its completion in 1978. The cryogenics maintains the coils of the magnetism a superconducting quantum state, and the magnets consume only 1/3 of the power they would be required with copper magnets at normal non-cryogenic temperatures.[23]

The Fermilab accelerator complex was much like the transmission in your car, having different "gears" that sequentially deliver more energy as you increase the car's speed. We began in first gear with a high-voltage Cockcroft–Walton pre-accelerator. This device was a large-voltage source that gave one big initial energy kick, very similar to that used in old TV picture tubes but on a much larger scale (TV picture tubes used a voltage of about 20,000 volts, while the Cockcroft–Walton at Fermilab produced 750,000 volts). The accelerated beam then passed into a 150-meter linac, accelerating up to 400 MeV, then into the "Booster," a small synchrotron about 100 meters in diameter. Here the protons circulate in their orbit about

20,000 times and attain an energy of around 8 GeV. From the Booster the particles pass into another synchrotron called the Main Injector, which accelerates the protons up to 120 GeV. Some of the protons at this stage were used to create antiprotons, which were collected into a sophisticated device called the antiproton source.[24] This made the antiprotons available to be injected back into the Main Injector, and the antiproton energy was increased back to 120 GeV. Finally the protons and antiprotons were both injected into the Tevatron.

The Tevatron could accelerate the protons and anti-protons from the Main Injector in opposite directions up to 980 GeV each. To maintain the particles in their synchrotron orbits the Tevatron used 774 niobium-titanium superconducting dipole magnets, cooled to superconducting temperatures in liquid helium. 240 quadrupole magnets were used as magnetic lenses to focus the beam.

The protons and antiproton beams were rather diffuse and generally didn't interact, passing freely through one another throughout most of the length of the circumference of the Tevatron. However, at certain special points around the Tevatron the beams were "squeezed" together and collisions occurred. This is the basic principle of a collider—the beams are not hitting a fixed target like a glass slide with protozoans in a drop of water. Rather, the beams are colliding head-on with one another! Surrounding these special "squeeze points" were the detectors, which collect and measure the products of the collisions. At the Tevatron there were two such detectors, CDF and D-Zero, electronically collecting and charting the debris of trillions of proton–antiproton collisions at 1.96 TeV.

The Main Injector, which replaced the old Main Ring of the Tevatron, was the last addition to the Fermilab complex, built at a cost of $290 million. The Tevatron collider Run II began on March 1, 2001, after the completion of the Main Injector, with the beam energy of 980 GeV. The Main Injector was the last particle accelerator for high-energy physics built in the US—and it was begun over 20 years ago in 1993. The Main Injector remains operational as an important part of the ongoing Fermilab program investigating neutrinos.

The Tevatron ceased operations on September 30, 2011, and some of its components have now been cannibalized for other accelerators and experiments.

LARGE ELECTRON-POSITRON COLLIDER

The Large Electron–Positron Collider was built at CERN and began operation in 1989. It was a circular synchrotron collider with a circumference of 27 kilometers and was constructed in the tunnel that now houses the LHC.

The concrete-lined "LEP tunnel" was a major construction project, undertaken between 1983 and 1988. The tunnel crosses the border, underground, between Switzerland and France, with most of it lying under France. The tunnel is tilted, and the high part of the ring is under the Jura Mountains to the west of Geneva. The tunnel therefore has a variable depth ranging from about 160 to 570 feet. Construction had to overcome serious challenges posed by underground water at high pressures in the mountains.[25] Hydrostatic cement held the day!

LEP accelerated electrons (in bunches) in one direction and positrons (antielectrons) in the opposite within a common beam pipe. Each particle bunch initially reached a total energy of about 45.5 GeV, yielding a combined energy of 91 GeV. This allowed the direct production of the Z^0 boson, which has a mass of 91 GeV. LEP was later upgraded to go to higher energies, which enabled the production of a pair of W bosons, each having a mass of 80 GeV. The LEP collider energy eventually achieved a beam energy of 104 GeV, for a total collision energy of 209 GeV.

Like the Tevatron, the particle acceleration at LEP was done in stages. The older CERN Super Proton Synchrotron was used initially to accelerate and inject bunches of electrons and positrons into the LEP ring. Once the particle bunches were accelerated to the desired beam energy, an electron bunch heading one direction and a positron bunch heading the other direction were squeezed to cause head-on collisions within the particle detectors. When an electron and a positron collide, they can annihilate to make a Z boson. The produced Z boson decays instantly into other elementary particles, which are then detected by the particle detectors.

The LEP collider had four detectors, situated symmetrically around the synchrotron, where the bunches of particles were "squeezed" to produce collisions. The four detectors of LEP were called Aleph, Delphi, Opal, and L3. These detectors, slightly differing in their designs, yielded complementary information about the physics at LEP. The detectors were quite large, each about the size of a small house. They could measure the decay particles from the Z boson produced in the collision. The detectors allowed a reconstruction of the process that produced them. By performing complex statistical analyses of this data, physicists could infer the properties of the Z boson in great detail.

The beam energy of the LEP collider was so precisely monitored it could detect the motion of the French *Train de Grande Vitesse* (*TGV*) as it rolled though the French-Swiss countryside en route to Paris or Lyon from downtown Geneva. Physics-wise, by scanning the Z^0 boson, one could measure and "count" the number of decays of the Z^0 boson into undetectable (or "invisible") particles called neutrinos. This confirmed the result that there were only 3 kinds (or "flavors") of very light-mass neutrinos in nature. The precision measurements of the Z^0 boson mass and decay have provided a major constraint on the indirect quantum effects in the Standard Model. Together with the discovery of the top quark at the Tevatron, this analysis gave strong clues as to where the Higgs boson would be found.

Though the original goal of the LEP collider was to discover the Higgs boson, we now know that it was slightly too massive to be produced at LEP energies and that it would have to await the LHC. LEP nonetheless made definitive and precise measurements of the properties of the Z boson. By carefully calibrating the beam energy and scanning the Z boson by varying the energy of the LEP collider, it was possible to infer many details about how the Z boson decays. A slightly higher-energy "super-LEP" machine might be built one day to produce the Higgs (in electron-positron colliders the Higgs boson is produced together with the Z^0 boson and requires about 240 GeV and higher luminosity than LEP). CERN terminated LEP to make way for the LHC in 2000.

LHC

The Large Hadron Collider is the world's largest and highest-energy particle accelerator.[26] The collider is contained in the circular tunnel and has a circumference of 27 kilometers (17 miles); it was originally constructed for LEP. It fully recovered from its "helium incident" (aka "major magnet explosion" on September 19, 2008; see chapter 1) and began doing physics in November of 2009.

Recall that the Tevatron accelerated protons (of charge +) in one direction and antiprotons (of charge –), in the opposite direction. This could be accomplished in one beam pipe with a common set of magnets, since both bunches would be held in a common orbit circulating in opposite directions. However, the LHC collides protons head-on with protons. This spares the necessity of making antiprotons, but it requires two adjacent parallel beam pipes (and more complex magnets) since protons cannot be circulated in the same pipe in the same circles in opposite directions. The two beam pipes must also intersect to create collisions.

1,232 dipole magnets keep the beams on their circular paths, while an additional 392 quadrupole magnets keep the beams focused. In total, there are over 1,600 superconducting magnets made of copper-clad niobium-titanium that are kept at their operating temperature of 1.9 K (-271.25° C) by 96 tons of liquid helium. The LHC has now eclipsed the Tevatron in another aspect: it is the largest cryogenic facility in the world operating at liquid helium temperature. Most recently the LHC operated at a beam energy of 4 TeV, or 8 TeV in the total collision energy. This was sufficient to discover the Higgs boson but was short of the planned design energy.

As of this writing (March 2013), the CERN LHC is down for upgrades. It will come back online around January 2015 at the full design energy, whence the protons will each have 7 TeV, giving a total collision energy of 14 TeV. The increase in total collision energy from 8 TeV to 14 TeV causes almost twenty times as many gluon-gluon collisions that produce Higgs bosons and new particles, and will yield much more data. It will also nearly double the discovery reach for new particles that are unanticipated in the Standard Model.

The run of the LHC starting in January 2015 and extending through to around 2018 will be one of the most important voyages the human species has ever taken into an unknown wilderness. Only this run of the LHC with the large detectors—the experiments called ATLAS and CMS—has a chance of seeing if there is anything "beyond the Higgs boson."

THE DETECTORS

Of course, as in the case of microscopy, the art of particle physics is not simply a matter of accelerating and colliding beam particles. It involves the eyepiece of the microscope, i.e., how to see the debris of these collisions and how to detect any new and unexpected particles produced within the collisions. We have said very little about the behemoth detectors of particle physics in this book. Alas, the detectors of particle physics necessitate another whole book, and we must suffice to point you to the Internet for more. Just search on the keywords "ATLAS CERN" and "CMS CERN," and you're on your way. You'll find the web addresses http://atlas.ch/ and http://cms.web.cern.ch/; you can see the remarkably short URLs these two experiments enjoy, perhaps in part because CERN was the origin of the World Wide Web in the era of Tim Berners-Lee. (You can also search for other detectors such as "CDF" and "D-Zero," formerly at Fermilab, and the LEP detectors mentioned above.)

But we do want to take a moment to salute an old late friend and colleague, and a hero of the resistance movement in France during World War II: Georges Charpak.[27]

In the 1960s through 1970s, particle detection involved mainly taking photographs of collisions within large and cumbersome devices called "bubble chambers," followed by the examination of these photographs by a human eyeball. This is a slow, non-automated, and labor- intensive method. It could not possibly deal with the very high statistics demanded of particle physics experiments that seek to find something like a Higgs boson, a needle in a haystack of trillions of collisions.

To make progress in detection, automation was required along with more advanced detectors. In 1968, the French physicist Georges Charpak developed the "multiwire proportional chamber." This is a gas-containing box with a large number of parallel wires, much like a piano harp, each connected to individual electronic amplifiers. Electrically charged particles passing through this array of wires ionize the gas and leave a small electric charge on the nearest wires. Amplifiers can turn this into an electrical signal, and this can be directly coupled to a computer. The system can automatically perform the task of particle detection at rates that are thousands of times greater than previously existing detectors.

Georges Charpak was later awarded the 1992 Nobel Prize in Physics for his work on the development of these automated particle detectors. Charpak's work has also significantly contributed to the use of this technology in many other fields that use ionizing radiation, such as biology, radiology, and nuclear medicine. Here again, we see knowledge driving the wealth of nations in real time.

BEYOND THE LHC?

The problem of designing higher-energy proton (and antiproton) colliders is solved: just build a bigger tunnel and a lot of magnets, etc., and scale up the LHC in size to whatever energy you desire.

Circular electron beam accelerators of ultra-high energy must overcome the problem of severe energy loss due to synchrotron radiation. This generally requires either a circular machine (a synchrotron) of extremely large radius or a return to the linear accelerator, but with a machine significantly longer and more expensive than those currently in use.

There is, at this writing, an active discussion under way to form a world collaboration with Japan for the construction of the International Linear Collider (ILC), which would be hosted on Japanese soil.[28] The ILC would essentially consist of two long lines of RF cavities. To make the Higgs boson we need to collide an electron head-on with an antielectron (positron) at a total energy of about 245 GeV. This production process makes a Higgs boson with a mass of about 125 GeV plus a Z^0 boson with a mass of 90 GeV. We would need another 30 GeV or so to make the collision rate for this process maximal, so we need a total energy of our electron plus positron to be about 125 + 90 + 30 = 245 GeV. To make collisions means we need one linac for the electrons of about 122.5 GeV and another for the positrons of 122.5 GeV, and then we must arrange to have the two beams collide head-on.

The proposed ILC is a machine approximately 22 kilometers in length, mainly due to the very long string of RF cavities required for acceleration. There are also a lot of additional systems needed to prepare the beams of electrons and antielectrons, and a complex "final focus system" to bring the beams into collision.

Some of the challenges with the ILC are: (1) Each pulse must contain about a hundred trillion electrons and a hundred trillion positrons, out of which only one pair of electrons collides; the pulses are not re-circulated and re-collided many times, as in a synchrotron, so the system uses a lot of energy.

(2) The ILC system, due to power considerations, cannot be scaled upward to arbitrarily higher energies and is ultimately limited to a highest collision energy of about 1 TeV; thus far the LHC has given no hints of new physics in this energy range, so this can only serve as a Higgs factory. (3) The cost of constructing the required string of 8,000 RF cavities is very high, previously estimated by the DOE to be over $16 billion dollars fully loaded, but the number we keep hearing now is $7 billion, and we don't understand what changed the arithmetic over the past decade. In any case, this project will have to draw upon a lot of contributing government resources.

Yet another possibility would be to build a large circular e^+e^- Higgs factory. The energy loss per turn in an electron synchrotron goes as E^4/R, where E is the beam energy and R is the radius. The synchrotron energy loss rate rapidly rises with energy, E, for fixed radius. But we can make the machine larger to somewhat reduce the energy loss, i.e., increase R. A detailed estimate suggests that such a machine operating under ideal LEP-like conditions could be built with an 80-kilometer-circumference tunnel. The technical challenges with this option are not quite as severe as for a linear collider, mainly because we only need one RF station with perhaps 100 RF cavities that the electrons and positrons pass through many times, but the idea of an 80 kilometer-circumference ring is daunting—we've already had a bad experience going down that path toward the SSC.

Such a machine could ultimately be converted to a very high-energy hadron collider, the Very Large Hadron Collider (VLHC), with energies three to ten times those of the LHC. This would bring the "energy frontier" back to the US, if it were ever to be built here. It would, alternatively, be an ideal machine for a growing country to invest in to get into the business of particle physics. Indeed, the Gobi Desert in Asia, or Siberia might be ideal places for such a machine.

Another option is to find a particle that is much heavier than the electron so it doesn't lose energy to synchrotron radiation but that should also be point-like, unlike the spongy proton and more like the point-like electron, so many of the advantages of using electron beams are available. That leads us to the *Muon Collider*, which we discuss later. To make the Higgs boson directly in a head-on collision, without the slower associated process involving the Z^0 requires a Muon Collider and would take only 125 GeV in total muon + anti-muon energy. This has a number of other physics advantages.

Far and away the biggest issues and challenges for all Higgs factories will be "What will we gain from such a Higgs factory that we won't already have learned at the LHC?" The LHC will do a superb job on improving the detailed understanding of the Higgs boson—will it really be worth ten billion dollars to squeeze that sponge a little more once we have all the LHC data in hand? Or should we do something else? Shouldn't we simply upgrade the LHC? What if the LHC runs very long and hard and finds no new particles beyond the Higgs boson? Shouldn't we improve our capability to indirectly access higher-energy scales in nature? Moreover, if the all-important LHC run starting in 2015 GeV *does discover new physics* at still higher energies than the Higgs boson mass scale, will we really want to have all of our limited resources diverted into an expensive and relatively lower-energy Higgs factory? Shouldn't we wait for LHC at 14 TeV, then think about the next big collider way beyond the Higgs boson? It is simply too early to chart the future course of colliders until about 2017 or so.

We can also explore new high-energy physics scales using indirect methods. It's called Project X, and it's a pathway beyond to a Muon Collider. Project X is smaller, less expensive, yet offers enormous discovery potential, and is something we could begin to build now.

CHAPTER 9
RARE PROCESSES

PRELUDES TO PARTICLE PHYSICS

From our modern vantage point, one of a deep understanding of fundamental symmetries governing all forces and space and time, and a wealth of experimental results from accelerators that reach way down to distance scales of a billionth of a billionth of an inch (10^{-18} centimeters), it is hard to fathom what it was like in the earlier, quaint era, the dawn of particle physics.

Today, modern, ultra-fast electronics and computers, and the development of strong superconducting magnets and radio-frequency cavities, which led to the great particle colliders, are the essential enabling technologies of physics. Producing W^+ and W^- bosons, Z^0 bosons, and top quarks is now the bread and butter of the LHC. The newly discovered Higgs boson, too, will soon become a familiar landmark along the trail to the shortest distance scales in nature. The LHC experiments will map out this newly discovered realm in great detail. We're now contemplating "Higgs factories" and the push deeper into the details of what the Higgs boson really is, and what may lie beyond.

A hundred and twenty years ago none of this could have been imagined. No one had yet noticed the "weak interactions" that involve the W^+ and W^- bosons flickering into existence for a miniscule instant in time as a "quantum fluctuation," according to Heisenberg's uncertainty principle. The weak interaction processes are so rare that we could seemingly have flipped a switch and turned them off altogether and no one would notice a thing. Yes, the weak interactions had created all the matter out of which we are made and had triggered the supernova explosions that redistributed it throughout the galaxies, so that solar systems with earth-like planets, and life, could form. The weak interactions had stylized the universe for us, giving us our home and the materials that are our being. But that story was hidden, deeply and completely, out of sight.

In a sudden burst of major scientific breakthroughs at the end of the nineteenth century the faint hints of the "weak interactions" were first noticed by humans. New enabling technologies had come into existence in the late 1800s, the technologies of vacuum pumps, high-voltage coils, and "electrical discharge tubes," the chemistry of fluorescence and phosphorescence, and photography. Particle physics emerged from the discovery of X-rays and radioactivity in 1895, while accelerators were still a long way off. This burst of discoveries came from careful study and analysis of *very rare processes* that nature only displays to the most patient and meticulous observers—and often, they came as mere serendipity.

THE FIRST RAYS OF DAWN

Wilhelm Conrad Röntgen was a German scientist, born on March 27, 1845, who received a PhD in mechanical engineering from the University of Zurich in 1869. He began research in physics at the University of Strasbourg, then took on a number of academic positions at various universities in Europe. Röntgen had once accepted an appointment at Columbia University in New York City, even purchased his transatlantic tickets, but he changed his plan with the imminent outbreak of World War I. He remained at Munich and died on February 10, 1923.[1]

In 1895 Röntgen was investigating the properties of "cathode rays"—produced by high voltage as an "electrical discharge," like a big spark—that passed through an evacuated glass tube. The tubes, when internally coated with fluorescent material, much like a modern fluorescent light, gave off visible light when the high voltage was applied. The high voltage was generated by opening a circuit containing a large coil that had been "charged" with current from a battery. This is exactly how a spark-coil works in an automobile, stepping up the lowly 12 volts of the car battery to many thousands of volts to cause the spark that ignites the fuel mixture. (This is also the phenomenon that caused the catastrophic magnet explosion at CERN as the LHC was first ramping up.) A large coil can be made to discharge its current into a vacuum tube, producing the motion of "cathode rays" (soon to be understood by J. J. Thompson to be fundamental particles, called "electrons," ripped out of their atoms by the electric force).

Röntgen noticed that he was able to get some "rays" from the powerful electric current to apparently exit the tube itself. The rays caused a shimmering of light on a small cardboard screen on which he had painted a known fluorescent compound, barium platinocyanide, close to the tube. But were these rays simply visible light coming from the tube, or were they some stray "cathode rays"? There seemed to be nothing exotic going on here.

To understand this phenomenon, Röntgen completely covered the tube with a cardboard blind so that no visible light could emerge. He then allowed the coil to discharge into the tube, and at that moment he just happened to see an observable shimmer on a fluorescent screen that was serendipitously placed on a table at a distance of about 6 feet from the tube. This was too far from the tube to be "stray cathode rays" (electrons), and there was no light coming from the cloaked tube itself. What was causing the mysterious distant shimmer?

This is a gorgeous example of an "accidental discovery," since Röntgen was simply trying to establish that the system was light-tight. In so doing he discovered something invisible that was getting out and producing the faint fluorescent shimmering light some distance away. He found that the experiment could be reliably repeated and modified, and he could consistently establish that there were mysterious ghost-like penetrating "rays" emerging from the tube as it discharged (scientific observations, unlike a séance that calls back the departed souls only when the spiritual weather conditions are right, are always reproducible, or else they're false). This new type of "ray" that was emerging from the electrical activity in the tube could penetrate the materials that blocked the pathway. Röntgen named these determined escapees from the discharge tube "*X-rays.*"

Röntgen threw himself into to the detailed study of the penetrating power of his X-rays. Within a few weeks he had produced a ghostly photograph of the bones in his wife's hand and even saw his own skeleton as an X-ray shadow cast on a fluorescent card, to which he declared, "I have seen my own death!" Röntgen later discovered that ordinary lead was an effective barrier to X-rays.

Röntgen had single-handedly discovered a marvelous new phenomenon of nature and then quickly developed most of the ingredients of the first effective medical/dental-imaging technology. It is no surprise that Wilhelm Conrad Röntgen received the *first* Nobel Prize in Physics in 1901. Today we know that X-rays are a very high-energy, ultra-short-wavelength form of light—they are very energetic and invisible photons.

INSPIRATION

Inspired by Röntgen's discovery of X-rays, a French scientist, Henri Becquerel, began to rethink "phosphorescence." This is the stimulated emission of light from a material, following the material's exposure to an external source of light, where the resulting emitted light from the phosphorescent material is generally of a color different than that of the source light (for example, a clock dial that glows in the dark after the room lights are turned off is phosphorescent). Becquerel had reasoned that phosphorescent materials, such as uranium, a fairly common mineral that was found in a black, otherwise seemingly useless gravel-like material called "pitchblende," might actually be coaxed to emit the newly discovered X-rays after being exposed to a source of bright sunlight. He did an experiment of placing the sunlight-exposed pitchblende onto a photographic plate, developing it, and finding that the plate had become "fogged," indicating that the pitchblende was indeed emitting something. Becquerel had discovered that X-rays came from uranium salts found in pitchblende. However, his *raison d'être*, i.e., his initial hypothesis of phosphorescence, was wrong. Upon performing subsequent experiments, he found that the X-rays were *spontaneously coming from the uranium* and needed no sunlight exposure to stimulate their emission!

Becquerel had discovered natural "*radioactivity*." Working with two brilliant doctoral students, Marie Curie and her husband Pierre, Becquerel subsequently discovered radioactivity in other "heavy" elements, such as thorium, polonium, and radium (the latter two elements were actually discovered by the Curies in the course of this research in Becquerel's lab). Together these three scientists shared one of the early Nobel Prizes for their discoveries of radioactivity and the new elements. In these experiments, they had observed three types of "rays" emitted by radioactive substances.[2]

With one of these newfound rays, the so-called "beta rays" ("beta" is the Greek letter β), they had witnessed, unknowingly and for the first time, the *weak interactions*. We've told you this story because it is through the weak interactions that we have, today, uncovered the celebrated Higgs boson. The Higgs boson, in giving masses to the force carriers of the weak interaction, mainly the W^+ and W^- bosons, causes them to become "weak" and hard to detect. It is a rare quantum fluctuation that causes a beta ray to be emitted from a decaying atomic nucleus. Yet, our century-long ascent into the physical world of the weak interactions began with the first steps of Becquerel and the Curies, way back in 1896 with the discovery of radioactivity. What came next was the grandest revolution in our understanding of nature, the development of the quantum theory.[3]

RUTHERFORD'S RADIOACTIVITY

We've previously met one of the greatest experimental physicists of all time, the New Zealander Ernest Rutherford. We described his most famous work: the discovery of the atomic nucleus, a discovery that was of paramount importance to the development of the entire quantum theory. However, this work was performed several years after he had already received his Nobel Prize, a rarity in the history physics. So what had Rutherford received his Nobel Prize for if it wasn't for the discovery of the atomic nucleus?

In 1898 J. J. Thomson, Rutherford's mentor at Cambridge University in England, had arranged an academic post for him at McGill University in Montreal, Canada. Here Rutherford set up his lab and explored the hot new topic of radioactivity. He soon discovered the concept of radioactive "half-life."[4] The main business was understanding all the various "rays" that are emitted from substances displaying radioactivity. The situation was confusing and full of initial mistakes and false hypotheses by all the players—but it was finally sorted out.

Rutherford's initial hypothesis was that all the radioactivity "rays" were just X-rays. However, by using the much more radioactive elements of polonium and radium, discovered by the Curies, he was able to show that there were two different rays that were deflected in magnetic fields, and therefore that these must be electrically charged particles. One of these was a slow and fairly non-penetrating form, which he called "alpha rays." The alpha rays required a very strong magnetic field to deflect them, and later Rutherford would prove that alpha rays were actually the nuclei of helium atoms (i.e., helium atoms with no electrons). These are fragments that are produced when a very heavy nucleus falls apart into a lighter one, emitting alpha particles. Some of the other radiation was not bent at all in a magnetic field and was deeply penetrating through matter and must therefore be electrically neutral. Rutherford called these "gamma rays," which are very energetic photons, of even higher energy than X-rays.

But Rutherford also found another kind of "ray" that was easily bent in a magnetic field, which he called "beta rays." Becquerel had also identified the "beta rays," but Rutherford now found that they were a hundred times more penetrating of matter than the alpha rays. From the deflection of their motion in a magnetic field, the "charge-to-mass ratio" of beta rays could be determined, and it was found to be identical to that of the electron, discovered by J. J. Thompson a few years earlier.

Usually the emitted beta particles have the negative charge of the electron. But in some rare materials the emitted beta particle has a positive electric charge: that's the antielectron, or *positron*. So, we'll now take a brief side-excursion into the fascinating phenomenon of antimatter (and by the way, here, too, has emerged another billion-dollar industry—positron-emission tomography—better living through particle physics[5]).

ANTIMATTER

We have all seen and admired the famous equation $E = mc^2$ emblazoned upon T-shirts, opening graphics for TV shows like *The Twilight Zone*, corporate logos, commercial products, and countless *New Yorker* cartoons. "$E = mc^2$" has become the universal emblem for "smart" in our culture today.

But this isn't exactly what Einstein said. What Einstein *really* said, is that for a particle at rest

$$\mathbf{E^2 = m^2 c^4}$$

To get the energy, E, for the particle, we have to take the *square root* of both sides of this equation, and sure enough, we'll then get a solution: $E = mc^2$. So what is the difference?

First, please just bear in mind a simple mathematical fact that you may have forgotten since you took high school algebra: *every number has two square roots*. For example, the number 4 has the two square roots $\sqrt{4} = 2$ and $\sqrt{4} = -2$; the latter is "negative 2." Of course, we all know that $2 \times 2 = 4$, but we also know that $(-2) \times (-2) = 4$ (two negatives make a positive when you multiply them together). The "other" square root of a positive number is always a negative number (and even negative numbers have square roots, which leads to imaginary numbers, but we digress).

So, then, here's the puzzle: If the true equation is $E^2 = m^2c^4$, then how do we know that the energy, E, that we derive from Einstein's formula should be a positive number? Which square root is it? Positive or negative? How does nature know?

Suppose *negative-energy particles* exist. These particles would have a *negative* rest energy of $-mc^2$. If they move, their negative energy would become a still greater *negative* quantity, that is, they would *lose energy* as they accelerate, i.e., their energy becomes more and more negative as their *velocity increases*. In fact, in collisions their energy would become more negative, and after enough collisions, eventually, the negative-energy particles would have an infinitely negative energy. Such particles would continuously accelerate and fall down into an enormous sinkhole of negative infinite energy. The universe would be full of these negative infinite-energy oddball particles, constantly radiating energy as they fell deeper and deeper into the infinite negative-energy abyss.

In 1926, a young British genius named Paul Dirac sought an equation for the electron, one that would be consistent with Einstein's theory of special relativity, as the equations of the time were based only upon Newton's concepts of space and time and they only worked for slow electrons.[6] Dirac found a beautiful equation in the new quantum theory, combined with relativity, and he hoped that it would lead to $E = +mc^2$. But he soon encountered a problem: his equation indeed had the "correct" solutions representing electrons that have spin 1/2 and positive energy, i.e., we do get $E = +mc^2$. But for every positive energy solution there was also a negative energy solution with $E = -mc^2$. The negative-energy electrons should be just as prevalent in nature as the positive-energy ones.

According to Dirac's equation, the universe should be full of these negative-energy electrons. The universe would become a purgatory, eternally collapsing into a great sinkhole of negative energy. Dirac became frustrated as seemingly nothing could be done about this conundrum—Einstein's theory of relativity combined with quantum physics predicted negative-energy electrons. This would imply that ordinary atoms, even simple hydrogen atoms, indeed, all of ordinary matter, could not possibly be stable. The positive-energy electron with $E = mc^2$ could emit a few photons, adding up to an energy of $2mc^2$, and could become a negative-energy electron with $E = -mc^2$, and then begin its orbital descent, accelerating and radiating more photons into the abyss of infinite negative energy. The whole universe could not be stable if the negative energy states truly existed. The negative-energy electron solutions of Dirac's equation were now a prime headache for the baby quantum theory.

However, Dirac soon had a wild idea. Wolfgang Pauli had successfully, and brilliantly, explained the Periodic Table of the Elements with a "rule" that must be obeyed by all electrons, known as the Pauli exclusion principle. This principle says that *no two electrons can be put into exactly the same quantum state of motion at the same time*. That is, once an electron occupies a given quantum state of motion, like a quantum orbit in an atom—that *state is filled*. No more electrons can join in. Quantum states are like seats on an airplane—only one passenger per seat is allowed. This is more than a mere "ordinance" or "rule," and Pauli actually proved it mathematically to apply for all spin-1/2 particles.[7]

Dirac's idea was a straightforward extension of Pauli's exclusion principle: He hypothesized that *the vacuum itself is completely filled with electrons, occupying all of the negative energy states*. And, if all of the negative energy levels in the whole universe are already filled, then positive-energy electrons, such as in atoms, could not drop down into these quantum states—they would be *excluded* from doing so by the Pauli exclusion principle. The seats in the whole vacuum are all sold out! The vacuum is now stable because it is already filled up with negative-energy electrons.[8]

Dirac thought that this was the fix, but he soon realized that it was not the end of the story. It was now theoretically possible to "excite" the vacuum. This means that physicists could arrange a collision in which they could kick a negative-energy electron out of the vacuum, much like a fisherman pulls a deep-sea fish into his boat. Dirac realized that this process would leave behind a *hole in the vacuum*. The hole, however, would represent the *absence of a negative-energy electron*. This means that the hole *actually would have a positive energy*. However, the hole would also represent the *absence of a negative electrically charged electron*, and hence the hole would be a *positively charged particle* (see figure 9.33).

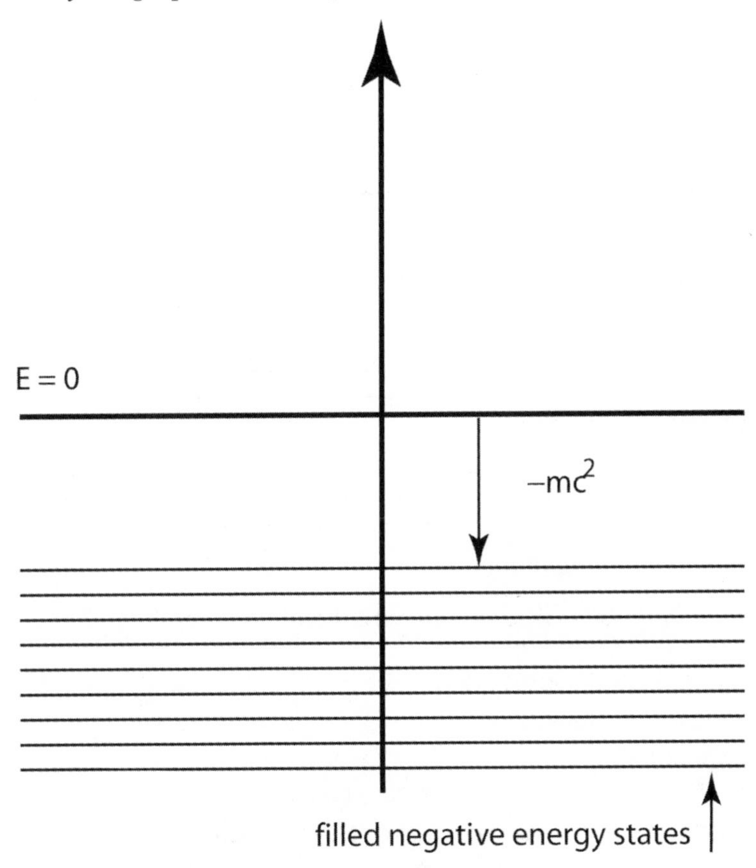

FIGURE 9.32. Dirac Sea. The "Dirac sea," picture of the vacuum. All of the allowed negative energy levels for fermions, predicted by Dirac's equation where relativity is combined with quantum theory, are filled. The Pauli exclusion principle forbids any more electrons in these levels, so the vacuum is "stable." The vacuum is like an inert element, e.g., neon, where all orbitals are filled, so neon becomes chemically inactive.

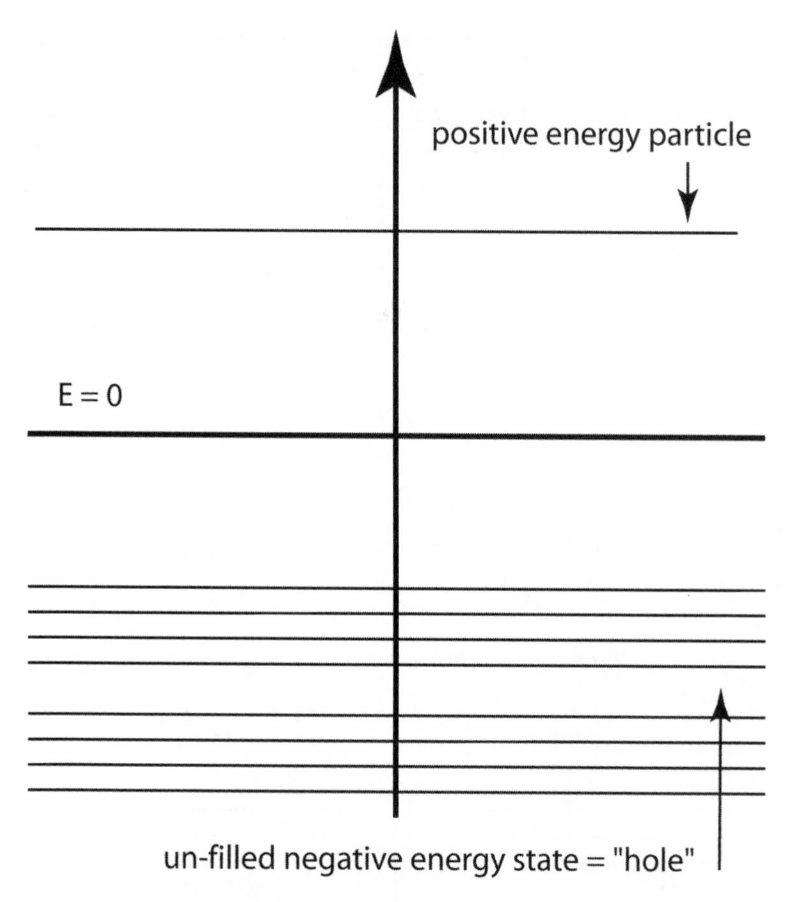

FIGURE 9.33. Antiparticle as Hole in Dirac Sea. Dirac's sea leads to the prediction that a negative-energy electron can be "ejected out of the vacuum," e.g., by the nearby collision. The hole left in the vacuum is the absence of a *negative-energy, negatively charged* electron, and therefore appears as a *positive-energy, positively charged* particle with identical mass to the electron. Dirac thus predicted the *positron*, and the phenomenon of "electron–positron pair creation." The positron was discovered experimentally a few years later by Carl Anderson. The phenomenon of antimatter is now well established and a standard phenomenon in particle physics—the Tevatron discovered the top quark by pair producing top and anti-top quarks in this way.

Dirac predicted the existence of something bizarre: *Antimatter.* Every particle species in nature has a corresponding antiparticle. We call the antiparticle of the electron the *positron.* The positron is the "absence" of a negatively charged electron, a hole, in the vacuum, and is therefore a positively charged particle with positive energy but is otherwise indistinguishable from the electron, with the same mass and the same (though opposite) spin. The laws of special relativity require that the hole in the vacuum, which is the *absence of negative* energy (note the double negative!), must have a positive energy of exactly $E = +mc^2$, where *m* is *exactly* the same as the *electron* mass. Positrons were predicted by Dirac, and they must exist if both quantum theory and special relativity are true.

Positrons were subsequently discovered in an experiment in 1933 by Carl Anderson.[9] They are the positively charged beta rays that are seen in radioactivity. Antimatter will annihilate matter when the two collide, as the positive-energy electron jumps back into the hole in the vacuum. The annihilation produces a lot of energy (at rest, electron–positron annihilation would release $E = 2mc^2$ by direct conversion of all the rest-mass energy of the two particles into gamma rays). Antimatter can easily be produced by particle accelerators.

Antimatter is a useful commodity and is already "paying rent." The positrons naturally generated from radioactive disintegration have found a use in positron-emission tomography (PET) scanners, a form of medical imaging. It is estimated that the cash-flow generated by this one activity, again a by-product of pure and basic research, is larger than the cost of funding all of the science of particle physics today. It is unclear if the future utility of synthesized antimatter will expand to warp-drive starship engines or compact super-energy storage devices, but eventually it will likely find many more practical applications—and yes, we're sure that one day the government will tax it.

Corresponding to *every* particle there is an antiparticle in nature. Corresponding to protons we have antiprotons, to neutrons we have antineutrons, to top quarks we have anti-top quarks. When we made top quarks in the good ol' days at the Fermilab Tevatron—now a staple of the CERN LHC—we made them in pairs: top plus anti-top. We literally go fishing and pull the negative-energy top quark out of the deep depths of the vacuum. This leaves behind a top quark hole (the anti-top quark), and we see the pair, top quark and anti-top quark, produced in our detectors. Particle physicists are simply metaphorical fishermen on the great Dirac sea.

BETA DECAY: THE SIMPLEST WEAK INTERACTION

The simplest example of the weak interaction is the *beta-decay* reaction that occurs with a single *neutron*, one of the particles found in the atomic nucleus, causing it to decay in about 11 minutes when it is floating about freely in space:

$$n^0 \rightarrow p^+ + e^- + \textbf{(missing energy)}$$

Beta decay is observed throughout many atomic nuclei, and it always involves this basic reaction. But beta decay posed a new problem: what is the "missing energy"? From countless observations, the electron and proton energies in the final state of the decay process always added up to something less than the original neutron energy. There thus appeared to be *a missing amount of energy* in the decay of a neutron. Essentially all beta decays of nuclei are a variation on this process, where the neutron is typically bound within the nucleus, and all revealed the mysterious "missing energy."[10]

Niels Bohr, one of the founding fathers of quantum mechanics, attempted to explain this phenomenon with the radical hypothesis that energy conservation, by which the initial energy is always equal to the final energy in any physical process, has only a limited validity in the world. Bohr proposed that the beta-decay processes were exhibiting, for the first time, a true violation of this time-honored and vaulted conservation law. Bohr, a brilliant and creative thinker, had already seen in the early part of the twentieth century that our detailed understanding of energy was significantly modified by the new rules of quantum mechanics, and he thought that perhaps beta decay was an indicator of deeper novelties and surprises yet to come.

Wolfgang Pauli, the brash and brilliant theoretical physicist who had developed his exclusion principle to explain how atoms with many electrons are built, could not accept Bohr's idea. The principle of the conservation of energy up to this point had proven valid in all domains of physics. It seemed unnatural to Pauli that the violations would show up *only in beta-*

decay reactions, where it is apparently seen to be a very large effect, and yet it doesn't show up elsewhere. Wouldn't any violation of this fundamental law of physics be universal, felt by all forces in nature, and not just be a property of beta decay? Bohr's proposal made no sense to Pauli.

In 1930, Pauli therefore did something quite radical by the intellectual standards of his day: he postulated the existence of a new and unseen elementary particle that was also produced, together with the proton and the electron, in the beta-decay reaction. This new particle must carry no electric charge and would therefore escape the decay region totally unobserved, and it would maintain the validity of the conservation law of energy, provided it had a very tiny mass. In other words, physicists could now compute the missing energy required to maintain the conservation law in any beta-decay reaction, and this would be the exact energy carried off by the new particle.

Pauli announced his new particle in a letter written on December 4, 1930, in a response to an invitation to attend a conference on radioactivity, which he declined.

> Dear Radioactive Ladies and Gentlemen,
>
> As the bearer of these lines, to whom I graciously ask you to listen, will explain to you in more detail, how because of the "wrong" statistics of the N and Li^6 nuclei and the continuous beta spectrum, I have hit upon a desperate remedy to save the . . . the law of conservation of energy. Namely, the possibility that there could exist . . . electrically neutral particles, that I wish to call [neutrinos], which have spin 1/2 and obey the exclusion principle . . . and in any event [have masses] not larger than 0.01 proton masses. The continuous beta spectrum would then become understandable by the assumption that in beta decay a (neutrino) is emitted in addition to the electron such that the sum of the energies of the neutron and the electron is constant . . .
>
> I agree that my remedy could seem incredible because one should have seen these (neutrinos) much earlier if they really exist. But only the one who dare can win and the difficult situation, due to the continuous structure of the beta spectrum, is lighted by a remark of my honoured predecessor, Mr Debye, who told me recently in Bruxelles: "Oh, It's well better not to think about this at all, like new taxes." From now on, every solution to the issue must be discussed. Thus, dear radioactive people, look and judge.

> Unfortunately, I cannot appear in Tubingen personally since I am indispensable here in Zurich because of a ball on the night of 6/7 December. With my best regards to you, and also to Mr Back.
>
> Your humble servant,
>
> W. Pauli[11]

The process of beta decay with Pauli's neutrino thus looks like this:

$$n^0 \longrightarrow p^+ + e^- + \overline{\nu}$$

The new particle, ν, is the *neutrino* (with a "bar" over the symbol, it becomes the conventional symbol for antiparticle, or *antineutrino*).[12] Therefore, when the neutron decays in free space, it produces a proton, an electron, and an (anti)neutrino. In our modern parlance, the electron is always produced together with the anti-electron-neutrino in a beta decay. The sums of the final energies of all of the three final particles will be exactly the same as the initial energy (mc^2) of the original parent neutron. Notice also that the neutrino, with zero electric charge, allows the beta-decay reaction to satisfy *the law of conservation of electric charge.* The zero electric charge of the neutrino means that it can't be easily detected—it lacks the "handle" of electric charge that we could otherwise "grab onto" through electromagnetic fields in our particle detectors.

With the details of beta decay now somewhat better understood, and Pauli's hypothesis of the neutrino, Enrico Fermi was able to write down the first mathematically descriptive quantum theory of the "weak interactions" in 1935. Fermi had to introduce a new fundamental constant into physics to specify the overall strength of the weak interactions, much like Newton had to introduce the "gravitational constant." In fact, Fermi's constant, called G_F, contains a fundamental unit of mass, which sets the scale of the weak forces—about 175 GeV. With Fermi's theory in hand it was now obvious that high-energy particle accelerators would eventually have to take over to study the details of the weak interactions. And not surprisingly, the Higgs boson has shown up with a mass of 126 GeV, not far from Fermi's "weak scale."[13] Pauli's neutrinos have also now been produced and subsequently detected in many experiments. Neutrinos were first directly detected by Clyde Cowan and Frederick Reines in 1956.[14] The neutrinos are important to us—they are emitted from the sun as part of the fusion process by which the sun shines. Our very existence depends critically upon the feeble weak interactions in nature.

THEY DID IT WITHOUT COLLIDERS?

Everything we have discussed to this point represents a fabric of some of the richest scientific discoveries in history—*and none of it was done with high-energy particle colliders!* Nature did all the work for us—it gave us unstable radioactive nuclei and cosmic rays as sources. The physics of beta decay led us to the discovery of the weak interactions and was all done in comparatively "low-energy" experiments, where nature furnished us with a "rare process." By patiently studying the details of matter, we could infer the deeper structure of nature that is the weak interactions.

Fermi's summation of the weak forces in his theory shows how the relevant high-energy scale of 175 GeV (or equivalent short-distance scale of about 1/10,000,000,000,000,000 centimeters) could be anticipated as early as 1935. It would be 25 years until Glashow would propose the W^+, W^-, and Z^0 bosons in the context of the symmetries of the Standard Model, and until Weinberg would show how a Higgs boson could break those symmetries. The actual structure of the weak interactions was not probed directly by colliders until the discovery of the W and Z bosons at CERN in 1985, and of the Higgs boson of 2012. So, from Becquerel to the Higgs boson discovery at CERN, the weak interactions have been at the core of physics research for almost 120 years.

What we have learned is that many of the great scientific advances in the twentieth century came by looking at the details of rare processes. Particle accelerators are generally either the colliders, operating at the highest energies, or smaller machines that produce many collisions, but each at a lower energy. Both offer pathways to discoveries.

For example, we now know that there are three different kinds, or "flavors," of neutrinos (see Appendix). Leon Lederman, Mel Schwartz, and Jack Steinberger demonstrated this in 1962 at the Brookhaven National Laboratory in Upton, New York, using a particle accelerator that provided a secondary beam of muons. With enough muon decays they showed that neutrinos are produced with distinct identities, by detecting the "*muon-neutrino*," which is a different particle than the "*electron-neutrino*." The key to the success of such an experiment is to have a very large number of particles available or a high statistics experiment. This required an *intense source* of sufficiently energetic protons to make muons, but not necessarily a very high-energy collider. With the subsequent discovery of the tau lepton, we now know there are three distinct "flavors" of neutrino."[15]

THE RARE WEAK PROCESSES

Let's isolate the process of beta decay and examine it in greater detail:

$$n^0 \longrightarrow p^+ + e^- + \overline{\nu}$$

At the level of quarks and leptons, the decay of the neutron, if viewed under an extremely powerful microscope, resolves into an individual "down" quark decaying to an "up" quark plus the emission of a W^- boson. However, the W^- boson is so heavy that this process can only happen by way of Heisenberg's uncertainty principle for a miniscule moment in time, allowing the energy to fluctuate by a large amount. The W^- then quickly converts into an electron plus a neutrino (see fig. 9.34). It is this extreme mass of the W^- that makes a weak interaction process very feeble, relying on the big quantum fluctuation in time and energy. The heaviness of the weak gauge boson is, in short, why the weak forces are weak.

This is the defining property of a "rare process." Rare processes involve something at very short distances over very short time intervals and occurring as a "quantum fluctuation." Becquerel couldn't observe the W^- boson directly in a radioactive beta decay, but by measuring beta decay in detail we could later theoretically infer its existence. With Fermi's theory we could actually infer the scale of weak interactions, of 175 GeV, and eventually write down a more precise theory, called the Standard Model, which told the future collider physicists exactly where to look. Physics is like the game "Clue" (also known as"Cluedo"). With enough indirect evidence from rare processes we could say, "It was Colonel Mustard in the Library with the lead pipe" who did it!"[16]

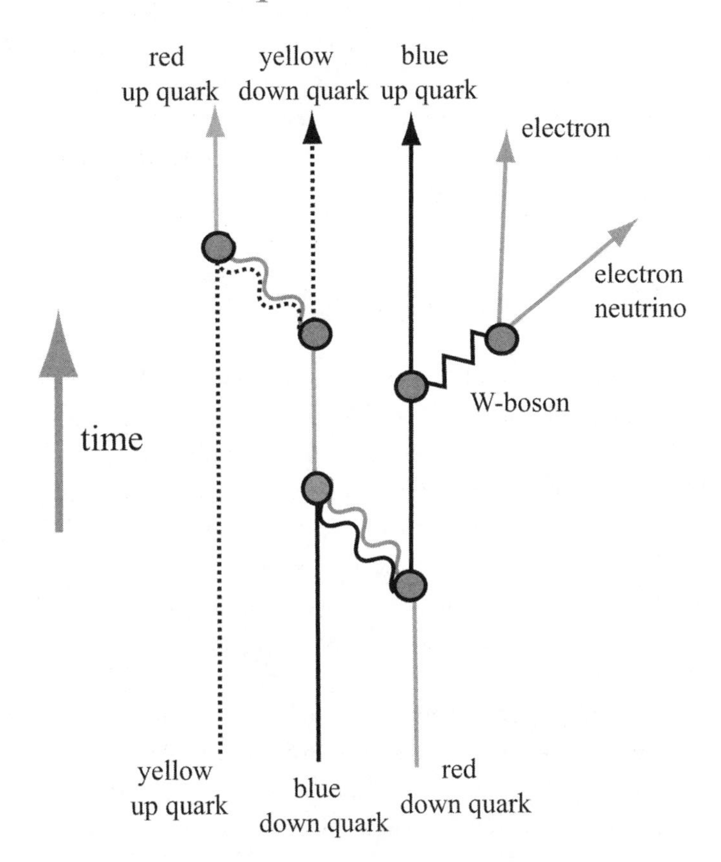

FIGURE 9.34. Beta Decay at Quark Level. At the level of quarks and leptons, and gluons binding the quarks inside the neutron and proton, we glimpse the process of a neutron decay, $n^0 \rightarrow p^+ + e^- + \overline{\nu}$, which involves the quark transition, $d \rightarrow u + e^- + \overline{\nu}^0$ through the exchange of a W boson. The W is so heavy that it is not produced as a real particle, but it is created for only a tiny instant of time (we call it a "virtual particle"), allowed in quantum theory by Heisenberg's uncertainty principle. The improbable quantum fluctuation makes the weak force very "weak." A free neutron has a half-life of about 11 minutes.

There are a vast number of rare processes that are part of the Standard Model. Many of these have not yet been seen yet in experiments because they are so difficult to observe. And many rare processes are suggested by theories that attempt to go beyond the Standard Model. These effects can show up in two ways, either as inconsistencies in the rates for certain rare process or as tiny effects that are forbidden by the Standard Model. What other tales might rare processes tell us about nature?

THE OTHER LOOKING GLASSES

As we've seen, until the mid-1950s, physicists believed that parity was an exact symmetry of physics. Thus, the world of Alice through the looking glass would be indistinguishable in any physical process from our own world. The question of parity (P) nonconservation in the weak interactions was first raised by two young theorists, T. D. Lee and C. N. Yang, in 1956.[17] Parity symmetry was practically considered to be an obvious fact in nature and had been used for decades in compiling data on nuclear and atomic physics. The breakthrough of Lee and Yang was the idea that the reflection symmetry—parity—could be perfectly respected in most of the interactions that physicists encountered, such as the strong force that holds the atomic nucleus together and the electromagnetic forces together with gravity. But Lee and Yang proposed that the weak force, with its particular form of beta-decay radioactivity, might not possess this mirror symmetry.

Indeed, as we've seen, parity violation was discovered experimentally in 1957 by studying the pion and muon weak-interaction decay in detail. Independently, the effect was seen by Madame Chien-Shiung Wu, using another technique in nuclear weak-decay processes. This was astounding news—the weak processes are not invariant under the parity (P) symmetry operation. Later it was shown that the parity violation of the weak interactions is a property of the fact that only L-handed particles participate in weak interactions, while their R-handed brethren do not. This in turn ultimately mandates the existence of the Higgs boson so that particles can march L-R-L-R through space-time and thus acquire mass.

Let us think a little more about space and time symmetries. Think about viewing the laws of physics by watching a movie. Parity symmetry would say that you cannot tell if the movie was taken by the camera looking directly at the scene or by viewing the scene reflected though a mirror. For example, the weak decay of a pion or a muon would not appear to us to be consistent with our known laws of physics if the camera is viewing things as a reflection in a mirror. L is exchanged with R, and we would see only R particles participating in the weak interactions in the mirror, not our world, where only L particles do.

But now let's try something different. Let's run the film backward through the projector. This is easy to do with a DVD player nowadays by pressing the "reverse" button. We have all seen, with amusement, the pie fly off Uncle Bert's face or brick towers un-collapsing and jumping back into their original positions. Unlike the world viewed through the mirror, it seems very easy to proclaim that you're watching a film that is running backward through the projector. We can therefore imagine a new kind of Alice mirror, a kind of "time looking glass" or "time mirror," in which we are always viewing things moving backward in time. Alice would find the world through the time mirror apparently quite different than her own. We'll call it "T."

However, when we examine up close simple or fundamental systems in the time mirror, such as two billiard balls colliding on the pool table, it becomes harder to tell in which direction in time the movie is progressing. The motion we see, forward or backward, viewed through the time mirror, as two billiard balls approach and bounce off one another recoiling into different directions on the table, appears not much changed. It's hard to tell if we're looking through the time mirror, T, or not. The forward-in-time collision on our side of the time mirror seems to respect the same microscopic laws of motion as the backward-in-time collision on the other side of the time mirror. The microscopic laws of motion of simple systems are evidently the same, whether they are run forward or backward in time.[18]

This is called "time-reversal symmetry." But just as parity seemed to be a symmetry until we encountered the weak interactions, we might ask, "Is time reversal a fundamental symmetry of nature and therefore valid for the elementary particles?" Does the world through the time mirror have the same laws of physics as does ours? Or like parity, is it a broken symmetry?

The answer is that the weak interactions, which violate parity, also violate time-reversal invariance at a much weaker level. To see this, we need yet another mirror—the antiparticle mirror.

CPT

We have already noted that mirror symmetry, designated by "P" for parity, is not a valid symmetry when it comes to processes involving the weak forces. Furthermore, as we have seen, there exists yet another discrete symmetry operation, called "T," which reverses the flow of time, that is, we can replace $t \to -t$ in all of our physics equations, swap initial conditions with final ones, and get the same consistent results.[19]

Yet another symmetry now arises given the existence of antimatter: it consists of replacing all particles by antiparticles in any given reaction. This is called C or "charge conjugation." This symmetry would imply that there is an analogy to Alice's mirror called an antiparticle mirror. When Alice falls through her parlor mirror, P, she enters the world in which all parities are reversed (all "lefts" become "rights" and vice versa). When Alice falls through the "antiparticle mirror," C, all the particles of all matter are turned into antiparticles and vice versa. We have seen that the laws of physics are slightly different through the parlor mirror—parity is a broken symmetry. So, naturally we ask, are the laws of physics the same through the antiparticle mirror? For example, would antihydrogen, consisting of an antiproton and an antielectron (positron) have the same identical properties, e.g., energy levels, sizes of the electron orbitals, decay rates, and spectrum, as does the ordinary hydrogen atom?

If C is a valid symmetry, then an antiparticle must behave in every respect identically to its particle counterpart, provided we replace every particle by its antiparticle in any given process. But this makes no reference to the spins of the particles, which have to do with P. In the pion decay, $\pi^- \to \mu^- \overline{\nu}^0$, the produced muon always has positive helicity, i.e., it is always produced as an L particle (negative helicity) in the weak interaction, but its mass flips it into R (positive helicity) so the process can occur and conserve angular momentum. If we perform a C operation on this process, we get the antiparticle process, $\pi^+ \to \mu^+ \nu^0$, where all particles are now replaced by antiparticles, but the spins all stay the same (we went through the C mirror, not the P mirror). Therefore, the helicity of the anti-muon in the antiparticle process would still be positive, or R (spin is still aligned with direction of motion).

In 1957, shortly after the overthrow of P, the symmetry C was tested directly through experiment. When the experiment was performed, the helicity of the anti-muon in pion decay was *not R*, rather it was *found to be L.* Therefore, the symmetry C is also violated, together with P in weak interactions, such as the decay of pions and muons.

The reason is not hard to see if you remember Dirac's sea. You'll recall that the L part of the muon couples to the weak interactions, while R does not. But antiparticles are holes, representing the absence of negative-energy particles in the Dirac sea. So, we would expect that if L has –1 weak charge, a hole would have +1 weak charge. But a hole is *the absence of L and must therefore be R.* So, for antiparticles we would expect that the R anti-muon couples to W bosons while the L anti-muon (the absence of R, which had weak charge 0) does not. Don't worry if you find this a bit confusing—it is, and it requires some practice to get it straight, and maybe a Tylenol® afterward. But it turns out that it would be hard to have it otherwise; when we reverse particle with antiparticle we naturally reverse parities.

So, naturally, there arose the conjecture that, perhaps, if we simultaneously reflect in a parlor mirror, P, and then go through the antiparticle mirror, C, i.e., change particle to antiparticle, that this combined symmetry may be exact in nature. The combined symmetry operation is called "CP." Upon performing CP to the negatively charged *left-handed* muon, we get a positively charged *right-handed* anti-muon. In the pion decay, $\pi^+ \rightarrow \mu^+ \nu^0$, the produced muon is indeed left-handed, so CP has turned out to be a symmetry of the pion decay. We now seemed to have deeper symmetry, which connected space reflections with the identity of particle and antiparticle. In summary, CP symmetry says: Jump through Alice's parlor mirror, which reverses parities, P, then jump through the antiparticle mirror, which changes all particles to antiparticles, C (the order is immaterial, CP is equivalent to PC), and we seem to get back to a world equivalent to our own.

But—the world is often much more enigmatic than humans are led to believe. In 1964, in a beautiful and extremely well-executed experiment involving some other interesting particles called neutral K-mesons (again, this is an accelerator-based experiment where "intensity," or many, many produced K's, is more important than high energy), it was shown that CP is *not conserved*, that is, *CP is also not a symmetry*. The physics of weak forces is *not invariant* under the combined operations C and P. If you go through the P mirror and the C mirror, you do not come home, but rather you end up in a world with different properties than our own.[20]

The details of the origin of this breakdown of the symmetry, CP, has come to define a frontier of physics for the past 50 years. There remain many unanswered questions, such as "Do neutrinos in their peculiar flavor oscillations interactions also display a violation of CP symmetry?" (See the next chapter.) We still do not know how this will play out, but we have since learned that if CP were indeed a perfect symmetry of nature, our universe would be so totally different that we, our solar system, stars, and galaxies, would probably not exist. Nor would you be reading this book. So, it's a good thing for us that CP as a symmetry of nature is actually violated.

CP violation tells us that a particle and an antiparticle do behave in slightly different ways. In fact, CP violation is a prerequisite to explaining yet another enigmatic question: "Why does the universe seem to contain only matter and no antimatter?" If we go back to the initial instants of the big bang, when the universe was extremely hot (hotter than any energy scale ever probed in the lab), cosmological theory would predict equal abundances of matter and antimatter. However, with CP violation, some ultra-heavy-matter particles could have decayed slightly differently than their antiparticle counterparts. This miniscule asymmetry could have favored, at the end of the decay sequence, the production of a slight excess of the normal matter (hydrogen) over the antimatter (anti-hydrogen). Then, as the universe cooled, and all the remaining matter and antimatter annihilated each other, this slight mismatched excess of matter remained. The slight mismatched excess of matter is us and everything we see in the universe.

The problem is that, while we need CP violation to explain the fact that the universe contains matter and no antimatter, we don't think we have yet discovered *the particular CP-violating interactions* that produce this effect. The CP-violation effect, first seen in neutral K-mesons, now seen in other particle decays, remains an intriguing hint of much more to come, but it cannot explain the matter–antimatter asymmetry. This issue is being studied aggressively around the world. And the answer may ultimately come from the lowly neutrino, if indeed neutrinos display CP violation. The devil is in the details. We have reached the frontier—we don't know the answer to this question.

DOES ANY COMBINATION OF MIRRORS TAKE US HOME?

Alice now has three mirrors to jump though. There's her parlor mirror, which flipped parities, P. There's the time-reversed mirror, T, which runs things backward in time, and there's the antiparticle mirror, C, which flips all matter into antimatter. Is there a sequence of mirrors we can jump through that will get us back home to the same world in which we live?

Quantum mechanics makes probabilistic predictions for the outcome of events. When we flip a "fair" coin, we have equal probability of getting heads or tails. But even with an "unfair" coin, the sum of the probabilities of getting heads or tails in a coin flip is one—the sum of all probabilities that anything should happen must add to one, or else we are not able to talk meaningfully about probability—the quantum theory would fall apart if this were not so. What would it mean that the probability of heads in a coin flip is 2/3, while tails is also 2/3? How can the total probability be 4/3?

It turns out that it is a theoretically necessary condition in quantum mechanics that, if we want the total probability of all possible outcomes for a given process to add to one, then the combined operations of CPT must indeed be an exact symmetry. If we combine C, P, and T, at least at the present level of experimental sensitivity, we do appear to have an exact symmetry of the world, CPT. There has been no experimental evidence of CPT violation, and many people consider it to be very unlikely. So—if Alice jumps through the C mirror, then the P mirror, then the T mirror (in any order), she gets back home!

If CPT failed as a symmetry, then over time probability would not be conserved. This undermines the notion of probability in quantum theory, and we would have to significantly modify it. That is, the probability for anything to happen under any circumstances would either exceed or be less than one! Nevertheless, we must ask, if the violation of CPT were very, very tiny, would we have noticed? It is, after all, an experimental question.

Let's step back and reflect on the situation. There are many questions we do not have answers to. The devil is always in the details, and physics is an experimental science. And, if there's one thing we have should have learned from history by now: rare processes, processes that may probe up to a 100 times or more beyond the LHC, may lead us to new physics. New discoveries could radically change our entire view of nature may be lying just beyond our current reach.

CHAPTER 10

NEUTRINOS

According to our modern scientific version of "genesis," the universe emerged from a plasma of the elementary constituents of matter: quarks, leptons, gauge bosons, and perhaps many other hitherto undiscovered particles furiously swarming about at extreme temperatures and pressures in an embryonic warped and twisted space and time. Space itself exploded, driven by the raw energy of the constituents of the universe, as described by the equations of Einstein's general theory of relativity. As the universe and its constituent plasma expanded, it cooled and condensed, ultimately transforming itself into a uniform gas of hydrogen, some helium, and relic particles of electromagnetic radiation, neutrinos, and some unknown(s) that are referred to as "dark matter." Primordial quantum fluctuations in the density of these relic particles may have been transmitted, through gravity, to the hydrogen gas cloud, leading to its collapse, and the formation of the galaxies and the first "protostars" of the early universe. These monstrous stars were the parents of all the later heavy elements, the planets, and the solar systems to come, including our own sun.

All the atoms heavier than helium, such as carbon, oxygen, nitrogen, sulfur, silicon, iron, etc.—the stuff of our own solar system, rocks, and our solid and wet planet; the stuff of life itself—were created within the gigantic protostars. The heavy elements were cooked by the process of nuclear fusion, within their cores, bound by immense gravitation, deep within these super-massive stars. These heavy atoms became the raw ingredients of the modern universe, without which there would be no structure. Eventually, by the parentage of the protostars, the sun and planets formed, and the special conditions on Earth led to the subtle and gradual evolution of life and of human beings. The true scientific story of our heritage is richer than any fables, and it is more mysterious and bizarre in its reality.

In order for this to work, somehow the heavy elements must become liberated from the cores of the super-massive protostars in which they were formed. Indeed, the nuclear furnace interiors of these monstrous stars eventually poison themselves. Filling with iron, the most stable atomic nucleus, they can no longer burn by nuclear fusion. The protostars then begin to collapse. Commanded by gravity, they cave inward upon themselves. No longer opposing gravity with the intense radiation of their nuclear engines, a sudden and rapid change occurs deep within their cores. There, the atoms of iron, supporting the entire weight of the massive hulk against the collapse by gravity, like the hull of a sinking submarine, give way and implode. The iron atoms are squeezed, subject to enormous pressure and density. This instantaneously creates a new state of matter, never before present in the universe—solid neutrons.

We've come a long way from Democritus, and we've learned that atoms consist of *electrons*, outwardly orbiting the compact nucleus that defines the center of the atom. The nucleus is made of *protons* and *neutrons*. When a protostar reaches its last stage of collapse, the electrons and protons in its core are squeezed together, merging within one another. A new set of physical processes, normally silently lurking in the background shadows of the everyday world around us, suddenly jumps to the fore. These are the *weak interactions*, the lowly radioactive decays that were not previously observed until Henri Becquerel's work in the 1890s, and they quickly convert the squeezed protons and electrons into neutrons. This produces, as a by-product, an explosive blast outward of elementary particles, the *neutrinos*. The dominant process of the weak interactions that destroys the monstrous protostars takes the form:

$$p^{+} + e^{-} \rightarrow n^{0} + \nu_{e}$$

or, "proton plus electron converts to neutron plus electron-neutrino." It's just beta decay slightly rearranged.

At the instant of the collapse of the core of a protostar, the weak interactions have stolen the show. The innermost core of the star is compressed into a ball of pure neutron matter, extremely compact, perhaps only ten miles in diameter, and yet as massive as our sun, but a trillion times more dense. The neutrinos, however, stream frantically outward from the core. As the neutrinos burst forth, they exert extreme pressure on the dense, hulking outer parts of the protostar—*the outer shell of the star explodes.* This marks a *supernova*—the most intense and spectacular explosion to occur in the universe, after the big bang.[1]

It is remarkable and ironic that this ferocious "mother of all explosions" involves the lowly neutrino, an elementary particle that would seem otherwise to be the most inert and inconspicuous of all particles. Out blast the neutrinos, taking with them all of the outer matter of the star, and all of the newly synthesized elements, producing a brilliant flash of light, many thousands of times brighter than all of the stars shining within a single galaxy. The outer shell of the body of the protostar, containing all the elements from hydrogen to iron, is blown out into space, making a gigantic cloud, or "nebula" from which future and second-generation stars and solar systems (and us) will form. A dense spinning neutron star, or perhaps a black hole, is left behind. This is the tiny remnant of the pure neutron core of the protostar that was blown inward in the mighty supernova explosion, a few miles in diameter, spinning on its axis faster than once per second, but with a mass greater than that of our own sun.

Over time, the nebulae of gas and dust and debris, now containing the heavy elements—the cindered remains of the many deceased protostars in their violent fates—accumulated and encircled the galaxies. This gave the galaxies a new and grandiose shape: that of gossamer spirals with their outreaching and enveloping spiral arms. In the outer spirals of the galaxies were born the offspring of the protostars, the second generation of smaller, yellowish stars, like our sun, together with the comets, asteroids, moons, and planets. These were composed of the gas and the rocky and metallic remains of the protostars.[2]

The existence of everyday matter, the existence of the planets and the world we inhabit today, the existence of life and *our very existence* owes to the violent annihilation of these anonymous protostars that died in the ferocious oblivion of their supernovae, billions of years ago. All of our "everyday matter" was cooked together within these monstrous conflagrations. This process of heavy-element formation is ongoing throughout the universe even today. Many smaller large blue giant stars exist today, shining with the light of the fusion of almost pure hydrogen and helium, dwelling within the inner recesses at the centers of galaxies, detonating from time to time. In otherwise dim and distant galaxies millions of light-years away, the supernovae light them up for a moment, flashing in the dark, distant universe like fireflies in the night. And some stars within our own galaxy, and not too distant from Earth, perhaps the unstable and dying Eta Carinae (eta kar-in-i), will one day brighten our own sky with their cataclysmic finales.[3]

THE PARTICLES WITH THE SMALLEST MASSES

Neutrinos hold a certain fascination because they are so weakly coupled to matter that they are very hard to detect, particularly at low energies. They are only detectable through the weak interactions. There are more than a hundred trillion neutrinos passing through your body every second, mainly from the sun. The sun emits neutrinos copiously as they are associated with the nuclear fusion processes that generate sunshine and the synthesis of atomic elements. These neutrinos pass freely through the earth, so your neutrino bath is harmless and continuous, and doesn't depend (much) upon day or night, whether the sun is up or has set.

Neutrinos come in three types, or "flavors." These are called "electron neutrino," "muon neutrino," and "tau neutrino" (sometimes we just call them 1, 2, and 3). They are named for their closest relatives in the weak interactions, the electron, the muon, and the tau leptons. This pairing of charged lepton (e.g., the electron) with its neutrino (electron neutrino) is part of the symmetry of the Standard Model (see Appendix).

For many years it was thought that neutrinos were massless particles, that they only come in a left-handed variety (and their antiparticle would therefore be right-handed), and that they do not couple to the Higgs boson. If particles are massless, then there is no L-R-L-R march through space-time. Massless particles are either pure L or pure R and always travel at the speed of light. However, in the past few decades, we have learned that neutrinos do, in fact, have miniscule masses, but they are masses that are extremely hard to measure, and hence they are very feebly coupled to the Higgs boson. To this day the neutrino masses have been detected but not precisely measured. However, neutrinos also exhibit a dramatic phenomenon associated with their masses: they "oscillate," between their various flavors (electron, muon, or tau) as they propagate through space.[4]

The neutrino masses, just like those of the other leptons or quarks, involve the Higgs field filling all of space. As you have learned, this leads to a "forced march"—the familiar L-R-L-R—where now the L and R are two distinct neutrino "chiralities" of "left" (L) and "right" (R). Each time a neutrino takes a step in the march, however, it very slightly changes its "flavor" identity. That is, if we go through one complete cycle, L-R-L, an L muon neutrino will end up as mostly an L muon neutrino, but it will pick up a little bit of electron neutrino or tau neutrino. So after many such steps the identity of the original muon neutrino has changed, and it has accumulated a significant probability of becoming a muon neutrino or a tau neutrino. (Likewise, if we "launch" an electron neutrino or a tau neutrino it, too, will similarly change identity.)

It would be as if every time your pet hamster took a step in his hamster wheel, he acquired a miniscule quantum probability of being a mouse. After many rotations of the hamster wheel you might find that he had morphed completely into a mouse. After some more running, perhaps he morphs into a rat. No doubt, you would become quite curious about this phenomenon, as have physicists become fascinated with neutrinos.

And, of course, this raises various questions: Does a mouse also morph back into a hamster? What does a rat morph into, a hedge fund manager? (Sorry, we couldn't resist that one.) Are there possible morphs other than mice, rats, and hamsters? Do things work in reverse as they do forward in time (time-reversal invariance)? Do left-handed hamsters behave the same way as right-handed ones? You can easily see the large multiplicity of questions we are interested in with regard to neutrinos.

In studying the weak interactions, we encountered the L neutrino (or R antineutrino). Only the L neutrino "feels" the W bosons and participates in weak interactions. Only later, with very sensitive experiments did we encounter neutrino mass, and therefore the march of the L into the R neutrino through the interaction with the Higgs boson. But—wait a minute—for such a long time we thought neutrinos were massless, so we didn't ever have to worry about the R neutrinos. Now we find neutrinos have masses, so if we make an L neutrino in a weak interaction, then what is the R neutrino into which it steps in the mass march?

Here is a mystery involving neutrinos that we don't encounter with electrons, or muons, or quarks. It's a little tricky—we have to do some bookkeeping—but it isn't too hard, so hang in there. We've learned that the phenomenon of mass always requires an L-R-L-R march, whereby, e.g., an L electron converts to an R electron, which converts back to an L electron, and so on. The L and R electrons are two different particles that become independent of each other if we turn off the effects of mass (which is what happens as they approach the speed of light, from a stationary observer's perspective). Furthermore, both the L and R electrons must have the exact same electric charge of –1 since electric charge is conserved.[5]

Because of the foundational constraint—the conservation of electric charge—we wouldn't dare hypothesize that the R electron is just the anti–L electron, that is, the R positron in disguise. Indeed, the anti–L electron, or R positron, has right-handed chirality (its spin is counter-aligned with its velocity at the speed of light—it's the absence of L in the vacuum, hence R). But the R positron has an electric charge of +1, so it cannot participate in a L-R-L-R . . . march, since the electric charge would then oscillate in time: (–1) . . . (+1) . . . (–1) . . . (+1) . . . hopelessly violating the conservation of electric charge. No way! The L and R electrons, and also the L and R positrons, are all distinct particles—the lowly electron involves a total of 4 different particles, or 4 components; the same is true for the other charged leptons, muon and tau, and for quarks as well. This 4-component system of a spin-1/2 particle is called a "Dirac particle."

However, neutrinos are different than quarks or leptons—they are electrically neutral—they have *zero electric charge*. It therefore becomes thinkable that an L neutrino can actually flip into *its own* anti–L neutrino, which has an R chirality. The point is that for neutrinos, *because they have no electric charge*, the one-step L-R flip can, theoretically, also *flip (particle) into (antiparticle)*. It is as though the hamster at every step on his wheel flips into an anti-hamster and subsequently back again (and then there are also the mixings of flavor, the slight probability of flipping into an anti-mouse and an anti-rat and so on). A particle that flips into its antiparticle when it does the L-R step is called a "Majorana particle" (my-hor-AH-na). It's actually doing the "L-anti-L" step.[6]

This is *the neutrino mystery*: we don't know if the masses of neutrinos are of the "Dirac form" (requiring 4 distinct components: L, R, anti-L, anti-R) or the "Majorana form" (requiring only L and anti-L). There may indeed be an independent R neutrino, and also the anti–R neutrino. The L-R-L-R march involves the flipping of an L neutrino, which we can produce in a weak interaction, into a "sterile" neutrino, R, which we only see because of the mass. In that case, neutrino masses are just like those of charged particles, and they are then of the Dirac form. However, it is entirely possible, and in fact very likely from a theoretical perspective, that an L electron-neutrino flips into its own anti–L electron-neutrino—the antiparticle is then the R state, and the mass is of the Majorana form.

How can we tell if neutrinos have Dirac masses or Majorana masses? This is a really big question, one we hope to develop the tools to answer in the future. It may only be answered by seeing a particular, previously unseen, ultra-rare nuclear process called "neutrinoless double-beta decay." Unfortunately, for lack of space we have to send you to another source to read more about this phenomenon.[7]

Most theorists believe the *observed neutrino masses* will prove to be of the Majorana kind. The reason for this is a remarkable observation about neutrino masses in grand unified theories. In short, the idea is that at very high-energy scales, those of grand unification (about a trillion times beyond the LHC energy scale, or 10^{15} GeV), neutrinos are hypothesized to start out with the 4 ingredients of Dirac masses, that is, a distinct R neutrino exists for each flavor (with its corresponding anti–R particle). Such R neutrinos, however, would be "sterile," coupling only through the Higgs interaction (such as in figure 6.22) and through the gravitational interaction.

But then the sterile R neutrinos could experience extremely high-energy and extremely weird processes that the other L neutrinos cannot. The reason is that the W boson coupling of conventional L particles constrains them in many ways. For example, suppose a "mini–black hole" underwent a quantum fluctuation and came briefly into existence at a tiny distance a trillion times smaller than the scales we have ever probed at LHC (or smaller still). The black hole is like a fish in the ocean, like a big fat grouper, that appears on the scene and eats a little fish, then disappears into the undersea gloom. The mini–black hole would see a lowly R neutrino and swallow it—poof! The R neutrino is gone, and the grouper then swims away and disappears. But the mini–black hole cannot swallow the L neutrino and then simply disappear, because the L neutrino has weak charge—it couples to the W boson. The W boson is like a fishing line attached to the L neutrino—when the grouper swallows it, it is now caught and can't swim away . . . there's still weak charge in the grouper's belly forbidding him from just disappearing. So, he spits the L neutrino back out immediately: "Ouch. I don't want to mess with those fishing lines," says the grouper, and then he's gone.

The effect of this would be that the mini–black holes interfere with the mass-generation mechanism of the Higgs boson for neutrinos. All the sterile R and anti–R neutrinos can be eaten by mini–black holes in quantum fluctuations, which in physics parlance gives them effectively a very large Majorana mass. But the Higgs boson can still cause L to convert to R as a big quantum fluctuation. This then causes the L neutrinos to acquire a very tiny Majorana mass. These are the neutrinos we would see, with the fish lines of W bosons attached to them—the L and anti–L neutrinos (the ones with weak charges). So, what we believe we are seeing, here in the land of broken symmetries, are effectively Majorana masses among the ordinary L neutrinos that are produced in beta decays.

In fact, there's some real meat to this argument, and it actually predicted the observed scale of neutrino masses way back in the 1970s.[8] So, neutrino masses may have some deep secrets in store. Neutrinos really seem to be probing energy scales a trillion times beyond the LHC! Neutrino masses may already be one of our best indirect probes of the scale of the grand unification, 10^{15} GeV or so, and perhaps the quantum effects of gravity.

NEUTRINO CP VIOLATION

The neutrino "flavor oscillations" likely also include a new form of CP violation. This means that the marching step from L to R, from (particle) to (antiparticle for Majorana masses), is slightly different than the step from R to L, from (antiparticle) to (particle). In our hamster metaphor, it means that the probability of the hamster becoming a mouse or a rat in a complete oscillation cycle, L-R-L, may be slightly different than the probability for an anti-hamster becoming an anti-mouse or an anti-rat in an R-L-R cycle. Antineutrinos would oscillate through a cycle slightly differently than neutrinos. Neutrino CP violation is of enormous interest, and it may provide the mechanism by which the matter–antimatter asymmetry observed throughout the universe was generated.[9]

LONG BASE LINES

The "neutrino flavor oscillation" phenomenon is such a slight effect that it requires a great distance over which the neutrino must travel in order to produce an observable change in flavor. It's just the fact that one full cycle of L-R-L mostly preserves the identity of the original neutrino, with only a miniscule probability of changing identity—the hamster then needs many, many steps on the hamster wheel to change to a mouse. The idea of neutrino flavor oscillation was first put forward in 1957 by physicist Bruno Pontecorvo.[10]

The first experimental evidence of neutrino oscillation was seen by Ray Davis with his experiment in the Homestake Mine in South Dakota in the late 1960s.[11] He observed a deficit in the number of solar electron neutrinos arriving at Earth in comparison to the *theoretical prediction*. He had built a large detector that was deep underground in a mine shaft, shielded from cosmic rays. His detector was only sensitive to electron neutrinos, which are the only flavor expected from solar fusion processes.

Davis observed a *deficiency* in the measured signal of neutrinos. This deficiency was interpreted as electron neutrinos launched from the sun changing their identity into undetected muon neutrinos or tau neutrinos during the long transit distance from the sun to the earth. The trouble here was that one had to accept the theoretical solar calculations as a basis for interpreting the experiment—what if the sun isn't a "standard star" after all?

By 2001, neutrino flavor oscillations were conclusively identified as the source of the solar electron neutrino deficit (see note 4). Also, much larger underground detectors around the world observed a deficit in the number of muon neutrinos coming from muon decays that were produced by cosmic ray collisions in the upper atmosphere (cosmic rays have served particle physics very well indeed!). The enormous Super-Kamiokande detector in Japan provided its first definitive measurements of neutrino oscillations in 1998, using a baseline of the diameter of the earth. This propelled Japan into the forefront of neutrino physics.

This experiment could determine to high precision the arrival direction of electron neutrinos, and it could even observe those that were coming from the sun, upward through the earth at night (when the sun is below our feet shining on the opposite side of the earth, the detected neutrinos pass through the earth, and the neutrino direction was upward, it is, of course, "downward" at noon with the sun overhead). The Super-Kamiokande experiment detected a slight variation in the number of solar neutrinos over a day-night cycle. Since the electron neutrinos (almost) freely travel through the earth unimpeded, this could only be interpreted as a neutrino oscillation where an electron neutrino oscillated into some other kinds of neutrinos that weren't detected. This is called a "disappearance" experiment since we are detecting a deficit of the expected electron neutrinos. The leader of this effort, Masatoshi Koshiba, won the 2002 Nobel Prize in Physics for this work.[12]

The Super-Kamiokande project was mainly the problem of building an enormous particle detector. This was the world's largest human-made vat of ultra-pure water, instrumented with thousands of large glass phototubes to detect the light produced by neutrino interactions in the ultra-pure water. We should mention that this kind of effort is no less fraught with danger than building and operating very large particle accelerators like the LHC—indeed, it is subject to the same kinds of "Oh, $&^%" disasters:

> On November 12, 2001, about 6,600 of the photomultiplier tubes (costing about $3000 each) in the Super-Kamiokande detector imploded, apparently in a chain reaction or cascade failure, as the shock wave, from the concussion of each imploding tube cracked its neighbours. The detector was partially restored by redistributing the photomultiplier tubes which did not implode, and by adding protective acrylic shells that are hoped will prevent another chain reaction from recurring (Super-Kamiokande-II).[13]

PUTTING NEUTRINOS UNDER THE MICROSCOPE

All of the experiments we've described thus far used the sun or the cosmic rays as a source to generate a detectable signal of neutrinos. Clearly, it is desirable to have control over *the source* as well as the target in a lab experiment. Therefore, it was inevitable that neutrino experiments would move into the accelerator lab (or to use nuclear reactors as sources). However, we still require moderate to enormous distance scales for neutrinos to run in order to observe the charges in flavor. This has given rise to "long-baseline neutrino experiments," where neutrinos are made at a lab, like Fermilab in Illinois, and are detected a long distance away, such as in a deep underground mine in northern Minnesota.

The modern long-baseline experiments are after the precise details that are involved in neutrino mass oscillations, and the search for what has become the "holy grail" of the subject: the discovery of neutrino CP violation. A typical and very sensitive experiment of this type is under way at Fermilab at present, where we launch muon neutrinos from decaying pions produced by the accelerator and allow them to travel 500 miles underground (mostly under Wisconsin) and detect their conversion into electron neutrinos in an underground laboratory in northern Minnesota. The experiment is called "NOνA" (pronounced "nova"), and it seeks critical information of the values of the masses of the neutrinos that is a prelude to the actual discovery of CP violation. The NOνA website and related sites describe this in greater detail. Many labs around the world, even CERN, are contemplating future accelerator-based long-baseline experiments in neutrino physics. So, too, is Fermilab, and we hope to do it in a big way.

Fermilab is currently developing a next-generation neutrino experiment called LBNE (Long-Baseline Neutrino Experiment). The explicit mission of LBNE is to discover (or confirm) the existence of neutrino CP violation and to make precise measurements of neutrino properties. LBNE demands much from the Fermilab accelerator complex, which will be used to provide an intense beam of neutrinos. The intense neutrino beam will be sent from Fermilab, where it is produced, through the earth, to a distant detector that will be located in the Homestake Mine of South Dakota (where Ray Davis had placed the original experiment that first saw the effect of neutrino oscillations renamed the SURF laboratory). LBNE can establish definitively whether neutrino CP violation exists.

"Intensity" is the name of the game for neutrinos—maximum "proton power on target" to make lots of pions, which decay into muons and neutrinos, providing the source for the launched neutrinos. Here the energies of the individual protons are relatively low—typically 3 to 8 GeV—but we accelerate many of these, so the beam power is measured in "megawatts." Such powerful beams are required for many other scientific quests into the deep fabric of nature where the secondary particles are of interest. These beams may be composed of muons or neutrinos, both derived from pion decays, where the pions come from the original protons slamming into a target. Or we may wish to study copious quantities of particles called kaons, or even very heavy rare isotopes like radium, francium, or radon (thus giving a new meaning to heavy metal rock 'n' roll). The applications of future intense beams of particles are coming into focus in the field of elementary particle physics.

The existing NOνA project,[14] the future LBNE project, and, ultimately, the construction of a futuristic Neutrino Factory provide a powerful evolutionary program in neutrino physics. Such a program will be sensitive to surprises. There may be hidden and unexpected new phenomena in the realm of neutrinos, such as the existence of new neutrino species or new interactions that are not found in our "Horatio dream" of the Standard Model. Does the hamster morph into new species we have never seen before?

An important aspect of LBNE will be its versatile and massive distant underground detector. Unlike Super-Kamiokande, which was an enormous vat of water, this will be a vat full of pure liquid argon, a highly optically pure material that allows greater sensitivity in recording the light emitted from neutrinos that interact within the detector, permitting a superb suppression of unwanted "noise" from background events. The detector will be the world's largest application of liquid argon, weighing several tens of kilotons. Such massive detectors are crucial for collecting sufficient events from the weakly interacting arriving neutrinos over such long distances. Liquid argon detectors have not yet been realized on such large scales, but advanced detector technologies will allow for a rich physics program beyond the study of neutrinos.[15] This includes a high sensitivity search for processes predicted in many grand unified theories, such as the aforementioned neutrinoless double beta decay, and the iconic process known as *proton decay*, which indirectly probes an energy scale of order 10^{16} GeV (this is a thousand trillion times beyond the scale of the LHC). It also includes the search for neutrinos that may come from any chance supernova explosions within our galaxy or its neighbors.[16]

CHAPTER 11
PROJECT X

Our neighbors often ask, "So, what is the future of Fermilab?" Fermilab is the sole remaining single-purpose scientific laboratory dedicated to elementary particle physics in the Western Hemisphere. Fermilab no longer operates the Tevatron, which up to the time of the LHC was the world's most powerful particle accelerator. The Tevatron discovered the top quark, and in its last days it spotted the Higgs boson in a unique decay mode and production mode that only the Tevatron could explore. Alas, for funding reasons, and the impact on other planned projects, it was terminated on September 30, 2011.

> Scientists first stopped the CDF and DZero detectors. They then stopped the data acquisition system and switched off the electricity to various sub-detector systems. Then they shut down the Tevatron. Helen Edwards, who was the lead scientist for the construction of the Tevatron in the 1980s, terminated the final store in the Tevatron by pressing a button that activated a set of magnets that steered the beam into the metal target. Edwards then pushed a second button to power off the magnets that guided beams through the Tevatron ring for 28 years. For about a week following the shutdown, accelerator operations worked to warm up the superconducting magnets, normally kept at 4.8 Kelvin. Once the magnets reached room temperature, crews began removing the Tevatron's cooling fluids and gases. It took about a month to fully shut down the CDF detector. Shutting down the DZero detector took longer, since the collaboration took data using cosmic rays as a way to double-check the calibration of its detector. The DZero detector was completely shut down after about three months.[1]

The termination of the Tevatron program marked the end of Fermilab's reign as "king of the energy frontier," since the Main Ring accelerator was first turned on in the 1970s. Unfortunately, this has given rise in the press to a false perception that the laboratory no longer has a mission in particle physics, and that its future has now become uncertain. But, in terms of future plans, the laboratory has many. Fermilab's director for Project X exclaimed to a reporter:

> "We have 10 accelerators here on site," says Fermilab physicist Steve Holmes, with the merest hint of irritation. "We turned one of them off, okay?" Like several scientists I spoke to, Holmes was keen to point out that colliding high-energy beams of particles is not the only way of discovering new physics with accelerators."[2]

FERMILAB'S PROJECT X

Fermilab has a unique and critical mission to find new ways to penetrate deeper into the fabric of nature. And, yes, we do have plans for another approach. It's a departure from the conventional "energy frontier" effort using particle colliders such as the LHC. It is complementary to the LHC. It marks a revival of an older approach—the very manner in which the science of new forces and the structure of matter at short distances began. It follows the lessons of the heroes of the grand generation, the pioneers of modern physics and discoverers of radioactivity: Henri Becquerel, Marie and Pierre Curie, Ernest Rutherford, and many others. They deeply probed inside of matter, to discover and study *ultra-rare processes* that ultimately revealed new physics.

Such an approach may reveal the first real chinks in the armor of the Standard Model. The energy scales (the short distances) that can be probed by this indirect route are hundreds to thousands of times greater (smaller) than those of the direct approach of a collider. When the style of the old physics of this pioneering generation of scientists is combined with the recent advances in the technology of accelerators and detectors, astonishing new opportunities abound. And the usual benefits to society of developing these new technologies—the "exogenous inputs" to the economy—will accrue.

As Fermilab evolves the Long-Baseline Neutrino Experiment (LBNE), which will ultimately aim a neutrino beam at the Homestake Mine in South Dakota, it is preparing in parallel for the eventual construction of the world's most intense particle accelerator: "Project X." Project X will be the centerpiece of the future of Fermilab and the US High Energy Physics program. Project X is a *high-intensity* proton accelerator, sometimes called a "proton driver." Incidentally, this has the mysterious name "Project X" not because it is shrouded in some kind of secrecy but simply because no one has come up with a better one. If you have any suggestions for a better name for Project X, please don't hesitate to contact us.

Let's start with something simple: there is a profound difference between "intensity" and "energy." At the LHC we have fewer protons in the beam, but each has the highest energy to which we have ever accelerated protons. At Project X we will have lower-energy protons in the beam (from about 3 to 8 GeV) but many, many more of them so that the overall beam power is the highest ever achieved.[3] In our microscope analogy it's like turning up the brightness of the particle beam and at the same time studying many different and exotic samples under the microscope to search for something new.

Project X is an ambitious and aggressive technological goal: The construction of about a *5-megawatt* proton accelerator with an energy of 3–8 GeV per proton. Project X would become a new enabling technology of much of the mid- to long-term research goals at Fermilab, much like the gas discharge tube was for Röntgen, or photo emulsions and phosphorescence were for Becquerel. Project X would give the US a powerful new scientific instrument to advance basic research.

The physics program with Project X is extraordinarily rich. Detailed studies of neutrinos at LBNE, which would require 30 years without Project X, can be done within a decade with Project X. The rarest decays of K-mesons, which first taught us about CP violation, become possible and may reveal new physics at energy scales approaching 1,000 TeV. Project X will open an entirely new probe of CP-violation (or, the "time mirror" in our Alice metaphor) physics by permitting the study of super-heavy atomic isotopes that may provide unprecedented sensitivity to the detailed properties of electrons, neutrons, and nuclei themselves. This enables the greatest reach for possible discovery of the *electric dipole moment* of the electron (see below), directly giving us a new window on CP violation and possibly a new window on dark matter. Project X will also enable us to build a "Muon Storage Ring Neutrino Factory" that would provide an unprecedented source of *both* electron and muon neutrinos and that would give us the capability to study the neutrino's physical properties at the highest level of precision and to search for new physics. And, Project X sets the stage for perhaps the most exciting high-energy collider of all: the Muon Collider.

PROJECT X NEUTRINO EXPERIMENTS

Up to now, all neutrinos we study are the product of pion decays, since pions are easy to make in large numbers if you have a very high-power accelerator, such as Project X. Pions, when they decay, only yield muon neutrinos, and this limits the possible neutrino oscillation studies we can do. We would ultimately like to launch an electron neutrino underground on its way to the Homestake Mine in South Dakota to see what it morphs into (recall our description in the previous chapter of hamsters morphing into mice).

Muons decay into electrons, antielectron neutrinos and muon neutrinos. Therefore, if we were to capture the muons from pion decay, place them in a *racetrack-shaped "storage ring,"* where most of the muons decay in the straight sections, they would give us a powerful beam of antielectron neutrinos, as well as muon neutrinos (we could alternatively place antimuons in the storage ring to produce anti–muon neutrinos and electron neutrinos) A neutrino factory would allow, for the first time, the study of the neutrino oscillations of launched electron neutrinos in long-baseline experiments. It would be as though we could launch a mouse instead of a hamster and see what it morphs into over a long baseline trip.

Wait a minute—did we say "capture the muons" and put them in a storage ring? They only live for two millionths of a second, so are we sure that's what we meant? Yes, we're sure. This has been done, but at nowhere near the intensity scale required for a Neutrino Factory.[4] Smaller muon storage rings have operated since the early 1970s at CERN and at the Brookhaven National Lab. The latter's ring is moving to Fermilab and will be used to precisely measure the magnetic properties of the muon, known as the "g-2" experiment.[5] The goal of the Neutrino Factory is to significantly scale the size of the storage ring and to increase the intensity of the muon beam circulating in the ring. The muon storage ring could provide the ultimate Neutrino Factory.[6] It also gives us a great deal of "batting practice" for the eventual construction of the Muon Collider.[7]

Fermilab has acquired enormous experience in the burgeoning science of neutrinos and currently operates several major neutrino experiments. It is upgrading its accelerator complex to improve them. As we have seen in chapter 10, neutrino CP violation may be of profound importance, as it may play a key role in the generation of the matter–antimatter asymmetry observed throughout the universe.

RARE KAON PROCESSES AND CP VIOLATION

CP violation was first observed in experiments with "kaons." Kaons are strongly interacting particles that are composed of a light quark, an "up" or "down" (or anti-"up" or "down") quark with an anti-strange (or strange) quark. Of particular interest are the anti-down, strange or anti-strange, down states. These are called the neutral K-mesons, K^0, and $\bar{K}^0$ (see chapter 9, note 20).

The neutral kaons have long been known to "oscillate" between one another as they travel through space, and they are forerunners of the neutrino oscillations. The detailed study of these kaon oscillations led to the original discovery of CP violation in physics (the fact that the time mirror takes Alice to a different world, not her own). The detailed properties of neutral kaons may reveal the surprise of a small discrepancy with the Standard Model and indicate the presence of some new physics. There are also charged kaons consisting of (up, anti-strange) or (strange, anti-up) quarks whose decays are also potentially sensitive probes that may also reveal new physics.

Kaon experiments that study the decays of these particles with trillions of produced kaons would yield a very high level of precision in monitoring the rarest processes in the standard model. These rare processes typically involve two W bosons "flickering" into existence for miniscule instants of time as quantum fluctuations. At the same time, top quarks, and possible new particles, can flicker in the same fluctuation, yielding potentially surprising signals. By measuring these processes in detail we can reach a new level of sensitivity to possible new and unknown physics beyond the Standard Model.

Of particular interest are two ultra-rare processes that involve decays of kaons into pions and neutrinos. These are a $K^+ \rightarrow \pi^+ \nu$ ν-bar and $K^0 \rightarrow \pi^0$ ν ν-bar. The latter process has a very precisely calculated Standard Model rate, and any deviations from this would be evidence of new physics. To fully probe these requires experiments capable of detecting about 1,000 of these decays of both the charged and neutral kaons.

Future Project X–based kaon experiments will be able to probe for new physics with unprecedented precision, up to energy scales of hundreds of thousands of TeV, well beyond the reach of any foreseeable high-energy colliders. Should a kaon experiment at Project X reveal a new rare process, it would be the direct analogue of the Becquerel-Curie discovery of the weak interactions of over a hundred years ago. It would provide a clear-cut goal for the next century of particle physics.[8]

The high-intensity proton beam of Project X would readily enable such experiments. The particular technology of the Project X accelerator design, called a "continuous-wave linac" (this means a continuous beam, rather than a more typical pulsed beam), would provide ideal conditions for these experiments, permitting major simplifications of the experimental apparatus. The measurements would reach the precision of a few percent for these extremely rare decay rates of the kaons, comparable to the uncertainty on the Standard Model prediction. This thus offers the ultimate sensitivity to any new physics in these processes that might alter the decay rates from their Standard Model predictions. The two experiments would additionally offer sensitivity to a variety of other rare kaon decays involving speculative exotic new particles.

RARE MUON PROCESSES: μ TO e CONVERSION

Rare decays of muons, such as $\mu \rightarrow e\ \gamma$, if observed, would also be a harbinger of new non–Standard Model physics, since this process does not otherwise occur at an observable level in the Standard Model. Some exotic theories predict decay rates for this process that could be within reach of experiment. A related process involves the conversion of a muon to an electron upon scattering off of an atom, known as "μ-to-e conversion." This could also be sensitive to exotic new physics mass scales that may lie at thousands of TeV.

Fermilab is planning an experiment designated "Mu2e," which will use the existing Fermilab 8 GeV proton beam from the old Tevatron Booster to search for the μ-to-e conversion process with sensitivity at a level 10,000 times better than previous experiments. This experiment will probe for new physics mass scales up to 10,000 TeV, significantly beyond the reach of the LHC. Project X offers the possibility of increasing the beam power to the experiment by more than a factor of 10, allowing an ultimate sensitivity ten times greater. If the Mu2e experiment discovered these exotic processes, the Project X–era experiment would offer the unique capability of distinguishing the underlying new physics by measuring the μ-to-e conversion rate using different nuclear targets.

If we're fortunate and a new process such as μ-to-e conversion is detected, we would again be dealing with the direct analogue of the Becquerel discovery of radioactivity in the 1890s. There would be much to do as a follow-on study, as this would likely be the first hint of a new force in nature, probably involving some new "X" boson that would become the bread and butter of a futuristic collider of the late twenty-first and early twenty-second centuries.

PROJECT X PROBES OF ELECTRIC DIPOLE MOMENTS USING RARE ISOTOPES

Electrons define the entire world of chemistry and biology, the world that our eyes can see and about which our brains can think. Electrons are essentially "us."

J. J. Thompson, in 1897, had discovered that the "cathode rays," comprising the electrical current in a gas discharge tube (sort of the precursor to the fluorescent lightbulb), were actually particles streaming through the tube. These particles had a very small mass compared to the atom (the rest of the atom's mass being the heavy nucleus), and Thompson showed that they were part of every atom in nature. These particles had a definite fingerprint: their ratio of electric charge to mass, or "e/m," is a definite value, which can easily be measured by bending their trajectories in a magnetic field. When Rutherford later studied the beta-decay radioactivity in detail, he found the emitted particles were electrons, because they had the same charge-to-mass ratio as Thompson's electron.

Today we know more about the electron and its interaction with the photon than of any other physical system in nature. This is all codified in the magnificent theory called "quantum electrodynamics," which was ultimately developed into a consistent, calculable theory of every aspect of electron-photon physics in 1949, due to Julian Schwinger, Richard Feynman, and Sin-Itiro Tomonaga, who shared the 1965 Nobel Prize for their efforts.[9]

We know that electrons, because of their charge and their spin, are each little magnets. A spinning electrically charged particle is, in effect, a loop of electrical current, and currents produce magnetic fields. This is called a "magnetic dipole" field because it has the same form as a bar magnet, which has two poles, N and S. An electric charge, on the other hand, produces an outwardly directed electric field, called a "monopole field." The equations of electricity and magnetism are such, as far as we know, that there are no magnetic monopoles in nature. But this leaves one final loophole: are there elementary particles that have electric dipoles? Does the spinning electron that produces a magnetic dipole field also produce a similar electric dipole field?

Let's return for a moment to Alice's parlor. Consider an electric field emanating from an electric charge (an electric monopole field) and a magnetic field emanating from a current loop. If we look at a mirror image of this, we see there is a difference in how electric and magnetic charges reflect in mirrors. If the electric field is emanating "outward" from the charge (a positive electric charge), then its mirror image will also show an electric field emanating "outward" from the charge—there's no change in the mirror image. On the other hand, if the magnetic field is emanating upward vertically out of the current loop, then in the mirror the field will reverse, emanating downward out of the loop. This can be understood by considering the mirror image of the current—if the loop plane is perpendicular to the plane of the mirror, then the current direction is seen to be reversed in the mirror, and this causes the reversal of the magnetic field.[10]

Now, if instead of the parlor mirror, we use the antimatter mirror, C, and we change the sign of the electric charge (that is, replace electrons with positrons), then both the magnetic field and electric field change direction.

So, if we do both a reflection in the parlor mirror, P, and follow by a reflection in the C mirror, C (swap all particles for antiparticles), for a net CP reflection, then the electric field always behaves oppositely to the magnetic field. So, if an electron produces a particular electric dipole field, then there must be a violation of CP symmetry, since the alignment of the electric and magnetic dipole fields of the electron select a preferred side of the combined mirrors. CP violation is always interesting because, since the combination of passing through the CPT mirrors always gets us back home, it is therefore associated with the arrow of time in physics, i.e., passing through the time mirror, T, undoes the effect of passing through the combined mirrors, CP.

No one has ever discovered an electric dipole of a point-like elementary particle. Certain molecules, like water, are famous because they spontaneously form a bent configuration that does have an electric dipole field. However, this is associated with the complexity of the water molecule. The bent molecule is a snapshot that is far from its pure quantum ground state. In its pure quantum ground state of rotational spin, even a water molecule has no electric dipole field. But truly elementary particles have "intrinsic spin" and by their very identity are always in their pure ground state of spin. For them, the existence of an electric dipole field is always violation of CP symmetry.

The discovery of a nonzero electric dipole field, or "electric dipole moment" (EDM), as it is more professionally called, for any elementary particle, e.g., the electron, would be of historic significance and could indicate the existence of new CP-violating physics. It would surely win a Nobel Prize for its discoverers. The CP violation in the Standard Model can produce an infinitesimal electric dipole field that is too small to be seen in the current experiments (it's about 10^{-38} in units that are the electric charge of the electron multiplied by 1 cm, or e-cm). EDMs provide potentially remarkable sensitivity to new physics.[11] Indeed, the known Standard Model CP violation among quarks is also too far away to explain the creation of the matter–antimatter excess and yet exist, so there must be some other source of CP violation in the universe. The search for nonzero EDMs is an excellent way to probe nature to try to get hints, in the spirit of Becquerel and the Curies, about what the other unknown sources of CP violation may be.

There are many exciting experiments that attack the problem of electric dipole moments. We can describe only one interesting line of attack presently based upon the remarkable fact that a big atom, an atom with a very heavy nucleus, provides an "amplifier" for EDMs of electrons.[12] Indeed, there are some really heavy nuclei. The heaviest ones, above uranium, such as radon, radium, americium, and francium, are all radioactive, that is, they disintegrate, and some don't live very long at all. The number of protons in the nucleus is always denoted by Z. The effect of large Z atoms is to amplify the effect of the EDM by large factors that grow as Z^3.

Project X can yield large quantities of heavy short-lived isotopes, such as radon, radium, americium, and francium, to support precision searches for the electron EDM. These experiments could significantly improve the existing limits by a whopping factor of 100 to 1,000. We think the technologies acquired here will lead to multi-billion-dollar economies in the future as well.

RIDDING THE WORLD OF PLUTONIUM AND PROVIDING ETERNAL CLEAN ENERGY: ACCELERATOR-DRIVEN SUBCRITICAL REACTORS

Particle physics demands and drives the creation of leading-edge technologies, the capability of studying systems as small relative to the atom as a basketball is small relative to the earth. Society and global economies have greatly benefited from the development of the most powerful particle accelerators and detectors, i.e., the most powerful "microscopes" ever created by humans. The advent of such applications of these technologies, such as advanced medical imaging, the very effective proton therapy for cancer treatment,[13] massive data handling and computing, and the World Wide Web, has paid back many thousands of times what the investments originally cost and has played a major role in defining our modern world.

New ideas, which require Project X for testing and development, offer the prospect of virtually infinite and clean sources of energy and the means to "incinerate" nuclear waste. These are profound goals: the generation of safe, clean, and abundant electrical power through accelerator-driven thorium reactions, and the potential "incineration" of radioactive wastes from conventional nuclear power production.[14] They can be explored at the "proof of principle" level at Project X, with eventual implementation of these technologies elsewhere.

The issue of high radio-toxicity and the long lifetime of conventional spent nuclear fuel is a global challenge. Accelerator-driven systems, like Project X, can be used to transmute spent nuclear fuel, which would significantly reduce the lifetime and toxicity of nuclear waste. Accelerators can also produce net energy by inducing fission in lower-atomic-weight elements such as thorium (Th). These are called "accelerator-driven subcritical" reactors because they don't require a "critical mass" of radioactive fuel and can never lead to an event such as the Fukushima or Chernobyl core meltdowns.

ADS reactors would have many key advantages: (1) it is estimated that the abundance of conventional reactor fuel, ^{235}U, is limited to about 100 years of energy production at current global demand rates, while the ^{232}Th isotope is abundant and is estimated to be able to provide 10,000 to 100,000 years of available fuel; (2) using Th fuel eliminates the production of toxic long-lived heavy actinides (such as plutonium and americium) and significantly lessens production of long-lived nuclear waste; (3) use of Th limits the possibility of nuclear weapons proliferation; (4) accelerator-driven reactors run in the "subcritical mode" and would be relatively safer to operate, i.e., one can turn off the accelerator driver and the reactor will shut down. There are no "core meltdowns" to fear.

The government of India and the US are contemplating formal cooperation in these areas of nuclear energy research in conjunction with Project X. Project X could support R&D of (1) the development of techniques for the destruction of spent fuel from conventional nuclear reactors, and (2) the development of ADS systems for safe and abundant nuclear energy production. We do not envision a full-scale accelerator-driven nuclear reactor development program at the Fermilab site, but key elements for the future of safe ADS nuclear energy can be studied and developed with Project X. Developing ADS reactors is, to us, a "no-brainer." It should have happened "yesterday." It is now becoming urgent.

BEYOND PROJECT X: THE NEXT COLLIDER

Over the next decade, experiments at the Large Hadron Collider will continue to explore a new energy regime and uncover the details of the Higgs mechanism that distinguishes the weak interactions from electromagnetism. The answer appears to be the Standard Model Higgs boson, but there may be a more elaborate accomplice—perhaps a new form of physics—new forces of nature, new symmetries, new particles, or new intricacies of space and time. Highly sensitive experiments at Project X will study rare processes, such as neutrino oscillations, EDMs, and very weak transitions among different quark and lepton flavors, and will indirectly probe energies well beyond those explored directly at the LHC.

To prepare to capitalize on any discoveries of new physics from the LHC and/or Project X–based experiments, Fermilab scientists are exploring the feasibility of a multi-TeV Muon Collider. This could be the highest-energy collider for the next generation beyond the LHC. The Fermilab community is leading physics and detector studies to map out the physics potential of a Muon Collider in terms of the machine's energy and luminosity. These studies will provide details as to how experiments could be carried out at a Muon Collider.

A Muon Collider uses muons and anti-muons as the "beam and target." The muons are produced from the intense Project X proton beam. The Muon Collider could begin life as a "Higgs factory," providing the best determination of the Higgs boson mass and directly scanning the Higgs boson in a unique way that no other collider can. The main advantage of the Muon Collider, unlike an electron linear collider, is that it could be scaled upward to become a very high-energy collider, effectively probing energy scales ten times greater than those at LHC. Any deep questions left unanswered from the LHC may ultimately be addressable by the high-energy Muon Collider.

A multi-TeV Muon Collider has many potential advantages over electron colliders, most of which arise from the lack of synchrotron radiation emission by muons, due to the heaviness of the muon compared to the electron. This allows a compact circular design of a synchrotron with multi-pass acceleration and multi-pass collisions. This could make for a cost-effective approach to reaching high energies with point-like lepton beam particles. Also, the Muon Collider would have a very narrow and well-defined beam energy. These are things that proton and electron colliders do not have. Electron linear colliders at very high energies, greater than 1 TeV, simply consume too much power due to not having the advantage of multi-pass collisions in a circular machine, since electrons lose their energy to synchrotron radiation. There is no physical problem in principle with a Muon Collider energy scale approaching in excess of 10 TeV (the equivalent proton collider energy scale for the same energy of point-like quark and gluon collisions would be about 100 TeV).

Fermilab leads the national Muon Accelerator Program (MAP) aimed at developing and demonstrating the concepts and critical technologies required to produce, capture, condition, accelerate, and store intense beams of muons.[15] Critical technologies are under study, including conducting experiments to demonstrate "muon cooling" (necessary to make a refined beam of muons and anti-muons), the study of RF cavity performance in the presence of high magnetic fields required for muon cooling, and the study of very high-field solenoids. MAP is also conducting advanced studies of beam dynamics, simulations of the muon production, capture, cooling, acceleration, and collision processes. The initial application of these new technologies might be the construction of a Neutrino Factory based on a muon storage ring.

Fermilab's expertise in high-field superconducting magnets will also be critical to any future synchrotron, such as a Muon Collider or Very Large Hadron Collider (VLHC), which both benefit from magnets capable of achieving the highest possible fields. For example, one design for a Muon Collider requires enormous 50-tesla focusing solenoids, while a 40-TeV VLHC in the LHC tunnel would demand 25 to 30-tesla dipole fields. Such magnets could be based on high-temperature superconductors operating at low temperatures, where they can carry high currents in high magnetic fields. Fermilab is engaged in R&D leading to the construction of the first high-temperature superconductor-based magnets for future energy frontier accelerators.

Q: HOW TO BUILD A STARSHIP?
A: START AT THE BEGINNING

With Becquerel's discovery of radioactivity, the weak interactions were seen for the first time. The methodology was quite different than collider physics today. For Becquerel and the Curies, one began with pitchblende. In pitchblende, there is uranium, and the radioactive disintegration of the radium atom reveals the physics indirectly. By analyzing lots of pitchblende, one could observe very rare processes and classify them. This is, after all, how all science begins—observation of phenomena followed by classification.

The key to the search for any rare processes is to have a large quantity of data. The data can be collider data at the LHC, where the search is now on for the various decay modes of the Higgs boson and any particles beyond the Standard Model. Higgs factories will aim at even more copious and cleaner samples of Higgs bosons. But in a world of relatively low-energy physics there are ultra-rare processes that can be studied to probe the fundamental laws of physics, and that could reveal new and previously unanticipated forms of physics. This would provide the necessary arguments to build the next collider.

This is the quarry of Project X. It is the logic of a world in which there's only a Standard Model Higgs boson, but no evidence of anything else "nonstandard" at the LHC. We believe it is a "no-brainer" that now we begin to pursue the search for new and beyond-the-Standard-Model physics with the high sensitivity and diverse program afforded by Project X, and the eventual capability of a return to the energy frontier with a Muon Collider.

CHAPTER 12

BEYOND THE HIGGS BOSON

We have told you the story of the Higgs boson. We have tried to give you an idea about why it exists, based upon what we've learned about the nature of mass in the previous century. We've seen how the understanding of the basic concept of "mass," known only as the "quantity of matter" since the ancients, became more profound in the late twentieth century at the deepest level of the basic building blocks of nature, the elementary particles.

We have seen that the masses of quarks and leptons involve the interaction of two disparate and different massless particles, a left-handed particle that has a "weak charge," together with a right-handed particle that has zero weak charge. Mass is an "oscillation" between left and right. The interaction of left and right requires a new particle that also has the weak charge of the left-handed component to maintain the conservation law of charge. This is the Higgs boson. It is mandated by profound symmetries that are fundamental and immutable in nature.

The masses of particles are generated when the Higgs field develops a "field" in the vacuum, inferred from Fermi's theory to have a value of about 175 GeV. The Higgs field, like an enormous magnetic field, extends uniformly in all directions throughout all of space and time. The Higgs field is effectively a great reservoir, filling the vacuum with its weak charge. Into this reservoir a left-handed particle can discard its charge to become an uncharged right-handed particle; likewise, the right-handed uncharged particle can acquire the weak charge from the vacuum to become left-handed. This leads to the oscillation in time—left-right-left-right—for all quarks and leptons; this is the phenomenon of *mass*. And like ordinary electric and magnetic fields, whose particle constituents are photons—the particles of light, the quantum of the universal Higgs field that binds left and right is the Higgs boson.

On July 4 of 2012, the discovery of the Higgs boson was announced at the home of the world's largest particle accelerator, the Large Hadron Collider (LHC) at CERN in Geneva, Switzerland. The Higgs boson has weighed in with a mass of about 126 GeV.

CONFUSED ABOUT BIG SCIENCE

Our fellow citizens often get confused about what big science is trying to do, perhaps because of what we tell them, usually in the media. For example, all too often we hear that colliders are built "to discover extra dimensions," to "confirm string theory," "to discover supersymmetry." False! Colliders are built to uncover *whatever is happening* in nature at the shortest distances, and not to accommodate the agendas of various sects of theorists. Often we hear that colliders are built "to re-create the conditions in the early universe (the big bang)." There's some element of truth to that, but in fact colliders don't re-create the thermal plasma in the hot, dense early universe; if they did, we wouldn't see the remarkable phenomena of quark jets (see Appendix) or CP violation in our collider experiments.

That this is confusing and mixing messages is best illustrated by something that happened about the time the Superconducting Super Collider (SSC) was terminated. We recall, long ago (but we can't remember exactly where) hearing a radio interview with a nurse who had just exited one of the large hospitals in Houston after a long day at work. A microphone was suddenly thrust in her face, and she was asked by a radio reporter, "Tell us, what do you think about today's cancelation of the Super Collider?" The nurse paused for a moment then replied, "We already have one universe, so I don't see why we have to create another one." The problem is that when people are told in a public presentation about all the latest and hottest gee-whiz theoretical and cosmic things, they often ask at the end of the talk, "What is the practical benefit of this?" "Why should I pay for this?" "What good is this?"

In fact, it's all about the world's most powerful microscopes. We have learned, by doing the experiment over the years, that people seem instinctively to "get it" when we tell them this simple fact: particle physics is the exploration of the smallest things in the world with the most powerful microscopes we humans have ever built. The audience then asks intelligent questions, such as "How big is a quark?" or "What is the magnification power of the LHC compared to the Tevatron?" They start to think like physicists. People have an inherent notion that microscopes are useful and important to humanity—that these are powerful scientific instruments studying the tiniest things and not antecedents to weapons of mass destruction or the end of the universe. Microscopes, to our friends and neighbors, *are useful.* They never then ask, "Why should I pay for this with my tax dollars?" (It's true—we've done this experiment many times in our talks and colloquia!)

Through this book we wanted to tell it straight. We have focused to a large extent upon the accelerators that have been built, the world's most powerful microscopes, how we have peeled away the layers of the great onion of nature, and the machines that we contemplate for the future. In any case, we've veered away from the "theories" as much as possible because, nowadays, accelerators and experiments are few and expensive, while theories are plentiful and cheap. Science is ultimately about measurement and observation, not just pure mathematics and wild, non-falsifiable speculations.

Particle physics is really the ultimate "materials" science, the study of the shortest-distance scale, the fabric of all matter—even the very fabric of the vacuum that fills all of space and time. The job of the world's most powerful microscopes is to reveal the smallest structures in nature, to tweak them and call them out of the depths of their sea, so we can understand them and, perhaps, see how it all works. The essential question of particle physics is: "What is matter and how does it work?" This was the question Democritus first asked in a scientific manner over two millennia ago, and beyond the immediacy of the discovery of the Higgs boson, we still have a lot of unanswered questions and a long way to go to find the answer.

THE CONNECTIONS

To be sure, the science of particle physics is indeed connected to other sciences in glorious ways. Since it deals with the quantum attributes of matter, it is intimately, conceptually connected to the study of "condensed-matter physics," and the weird and otherworldly ways that matter can behave under certain circumstances. We have probably learned the most about the possibilities for our vacuum and its various excitations (that's what particles are—"excitations" of the vacuum) from "superconductors," systems made of lead, or niobium, or nickel, which are cooled down to a few degrees above absolute zero, at which point they have absolutely zero electrical resistance. Such systems are "toy" universes that can be made by hand and variously studied in the lab. There is a sort of Higgs boson–like excitation found in superconductors, and the physics of a superconductor parallels and predates the theory of the Higgs boson of particle physics.

Particle physics is also connected to the study of cosmology in a fundamental way. In fact, the major breakthroughs in particle physics, culminating in those of the Standard Model revolution of the 1970s, allowed us for the first time to understand the big bang. The great discoveries, such as the "gauge principle" shared by all forces in nature, allowed us to speculate about "grand unification" and led to the idea of "cosmic inflation" and canonized the field of cosmology.[1] Suddenly cosmology became respectable. The leading cosmologists are all particle physicists. This has a certain irony because cosmology uses telescopes to look at big things that are very far way, while particle physics uses the most potent microscopes and studies the smallest things that are right under our noses and, in fact, that are us!

Indeed, the early universe is a place dominated by very high-energy collisions among particles, way up to and beyond the energy reach of our most powerful accelerators. Particle accelerators therefore yield fundamental information that is essential to understand the early universe. And particle physicists also know that there is valuable information about the elementary particles to be gleaned from the fossil record of the universe, i.e., the stuff that's left over from the big bang.

Perhaps one of the most interesting open questions is the existence of a mysterious and unaccounted for form of matter, called "dark matter," permeating the universe that is unseen by light but is nonetheless indirectly inferred from its gravity. It surrounds galaxies and great clusters of galaxies way out in the universe. The bigger the cluster of galaxies, the more dark matter we infer is there by studying the motion of the visible galaxies in the clusters. We can indirectly "see" dark matter as it bends light by its own gravitation, making enormous cosmic lenses in the sky.

But, as of this writing, while there are more theories of dark matter than there are feral cats in Chicago, the particle that constitutes dark matter has not yet been produced and detected in a particle accelerator experiment—dark matter hasn't yet been seen under a microscope (and dark matter may be plural)!

Dark matter therefore remains a mysterious quarry of the two conjoined sciences of cosmology and particle physics of our present day. So, these two sciences—particle physics, the ultimate microscopy, and cosmology, the ultimate "telescopy"—very much overlap, as they did in the era of Hans and Zacharias Jannsen and of Galileo, as the optical microscope and telescope were developed side-by-side. These sciences are intimately connected and symbiotically benefit from one another. Dark matter definitely informs us that there are things out there that we do not yet understand and that go beyond the philosophy contained in our Standard Model. There definitely is something beyond the Standard Model and beyond the Higgs boson. And there are so many unanswered questions within the Standard Model that clearly some deeper organizing principle(s) lie beyond it.

In many ways, cosmology is like studying the fossil record of dinosaurs, learning what once existed and what questions such things may pose for the overall structure of particle physics. Cosmology is an essential subdiscipline of modern physics. However, if you want to study the detailed processes that define what we call active "life," you need to go into the biology lab and use electron microscopes. Likewise, to understand what the basic constituents of matter are, and what the forces that control them are, you need to build a powerful particle accelerator, like the LHC or Project X, eventually, perhaps, a Muon Collider.

THE UNHEALTHY WEALTHY STATE

The health and wealth of nations critically depends upon the activity of basic research, including the seemingly more abstract construction of powerful particle accelerators. It is a no-brainer that powerful and able governments should fund it, even at the seemingly enormous costs it demands. The fact is that a world-class particle collider, nowadays, will cost some multiple of \$10 billion. That multiple may be 1 ×, or 1.5 ×, or even 3 ×. But on the scale of government spending, and of the scale of the wealth of nations, this is almost a trivial expenditure. Yet the US Congress is showing little interest in healthy science funding. Europe, Japan, and China are forging on.

To get a sense of scale, the US Navy's new *Gerald R. Ford*–class (CVN-21) aircraft carriers cost about $15 billion for R&D and construction. These will replace the 10 *Nimitz*-class nuclear aircraft carriers the US Navy currently operates and that cost about $50 billion just for construction (nuclear reactors, operations, etc., drive the cost up a lot more).[2] Moreover, the US sits on top of an *estimated* total $200 trillion—that's $200,000 billion—of coal, gas, and oil.[3] The total assets of households and businesses in the US is about $200,000 billion = $200 trillion,[4] while the top 100 richest US citizens have a combined wealth of about $1,000 billion = $1 trillion.[5] Particle physics gave us the World Wide Web, which creates an annual revenue stream globally measured in tens of trillions of US dollars. Yet endless squabbles persist in Congress over a national debt of $17 trillion (at this writing) and a deficit of less than $1 trillion. Meanwhile, the economy and the American standard of living falters, and science wilts on the vine.

HOW DOES THE HIGGS BOSON GET ITS MASS?

The Higgs boson of the Standard Model does explain (though some may prefer to say "accommodate") the masses of quarks, the charged leptons, the neutrinos, and the W and Z bosons. But it *does not explain its own mass*, about 126 GeV. It is the Higgs boson mass that determines Fermi's scale in the Standard Model. But we're still in the dark about the origin of the Higgs boson mass.

Where does the Higgs boson mass itself come from? That question has now moved to the forefront of the unanswered questions we have "beyond the Higgs boson."

This is rather frustrating for a significant reason: our very successful theory of quarks and gluons and the strong interactions, known as "quantum chromodynamics" (QCD), emerged from a series of breakthroughs in 1974. Once it was understood, and the quarks and gluons were confirmed, the theory neatly explained where the *strong masses* come from (see the Appendix). These are the masses of a long list of particles found in the 1950s and 1960s, and most of the masses of the proton and neutron.

In fact, we should apologize for not telling you this fact earlier, but strong mass, through the proton and neutron masses, actually makes up most of the visible mass in the universe—the masses of stars, planets, and large clouds of dust and debris of supernovas seen through telescopes. Very little of this actually comes from the fundamental and relatively tiny masses of the up quark, down quark, and electron. Strong mass comes from the inherent mass scale found in QCD, and not from the Higgs boson!

But QCD explains the strong mass scale in a remarkable and beautiful way—*it is due to quantum mechanics itself.* QCD starts out at extremely short distances (high energy) as a *scale-invariant* theory—that means it has no inherent mass scale at the outset—and the coupling of gluons to quarks is very feeble. However, due to quantum effects, the coupling of gluons to quarks becomes stronger and stronger as we descend to lower energies, or to larger distances. Finally, at a certain energy scale, about 100 MeV, or equivalently, a distance scale of about 0.0000000000001 centimeters (that's 10^{-13} cm), this coupling strength becomes virtually infinite. This causes the quarks and gluons to form composite states—the protons, the neutrons, the pions, and all the other strongly interacting particles. The quarks and gluons are then "confined" and are never observable outside of a composite state in the laboratory. This mass scale of 100 MeV is determined by the quantum interactions themselves—nature creates strong mass through its own dynamics, essentially out of no mass! It has nothing to do with any other scale of the onion of physics.

This leads to a beautiful conjecture about mass: *all masses in nature are generated by quantum effects.*[6] That is, if we could somehow "turn off" quantum theory, somehow make Planck's constant go to zero, we would live in a world with no mass—the particle utopia we described in earlier chapters. This is exactly how the strong interactions, as described by QCD, work. It is a natural idea to extend this hypothesis, and it immediately implies that the weak interactions would work the same way.

Unfortunately, the discovery of a Standard Model Higgs boson seems to have no obvious correlation with this hypothesis. We see no clue, at the moment, as to how to solve the riddle of the Higgs boson mass itself in a manner such as QCD generates strong mass. Nature has consolidated all of the quark and lepton and W and Z boson masses into the Higgs boson field, but the Higgs boson remains a black box—it does not yet tell us anything deeper about the origin of the electroweak mass scale, or equivalently, about its own mass.

FINALE

The most important next step for our science is the LHC run, scheduled to begin sometime around January 1, 2015, yielding possible major and dramatic new physics results in 2017 or so. Hopefully the LHC, when it comes back online, will reveal new particles and new phenomena, and the next layer of the onion will finally come into view.

Without such a revelation, without new targets for future colliders, can we rationally ask our government for a multi-billion-dollar high-energy particle collider at this time? The answer may be: we shouldn't.[7] It may be irrational and irresponsible to do so given that we have no indications of what new physics to pursue with such a machine. It would be a costly shot in the dark. Rather, we must wait until 2017 and continue reliably slugging it out at the LHC, participating actively in future machine and detector upgrades. There's still lots to learn from the LHC.

However, here in the US, we have a golden opportunity to penetrate deeply into the fog of the highest energies with a different, cost-effective approach, the approach of Becquerel, the Curies, and Rutherford, back in the earliest era of our science. We can now roll up our sleeves and build a smaller, few-billion-dollar machine, called Project X. With Project X, as we have seen, we could simultaneously probe nature for indirect hints as to what lies at energy scales 100 to 1,000 times beyond the LHC, while also collaborating actively in the energy frontier effort at the LHC. Project X could help to solve major global challenges, such as ridding the world of plutonium and providing clean nuclear power, as well as yielding rich scientific discoveries. It may ultimately lead us to the next-generation particle collider, first with a relatively small Muon Collider Higgs factory, using the powerful Project X beam to provide the requisite muon source. Later we could upgrade to a multi-TeV Muon Collider to provide point-like probes of any interesting new targets at the highest energies. This approach is staged, economical, and sensible. This is a most sensible evolutionary program that would allow the full benefit of advanced-technology R&D

to provide much-needed "exogenous inputs" into our economy This, we believe, is our best pathway forward, beyond the Higgs boson.

Experiment will always be the ultimate arbiter, so long as it's science we're doing. So far, regarding the Higgs boson there's not a hint of new dynamics. While we all expected that a major revolution was coming to the science of elementary particle physics, immediately with the discoveries at the LHC few expected a single Standard Model Higgs boson. So far the major revolution hasn't happened.

So what does this imply for the future? What else remains to be understood that can be understood? What, perchance, is not dreamt of in our philosophies? What generates the Higgs boson mass? Has the LHC missed something? Surely, there have to be some clues somewhere. Or maybe we're just not being clever enough? Are we misunderstanding what nature is telling us? We're working on it.

Please stay tuned for the all-important LHC results in 2017 or so. And let's roll up our sleeves and get started on Project X!

APPENDIX

THE STRONG INTERACTIONS

By the mid-1960s, a vast array of *strongly interacting* particles was produced in many experiments at the many new accelerator labs. The number of new particles surpassed the number of atomic elements. Almost all of these various new particles were cousins of the proton, the neutron, and the pion—the components of the atomic nucleus. These particles were unstable, some having comparatively "long lifetimes" of a hundredth of a millionth to a tenth of a millionth of a billionth of a second (10^{-8} to 10^{-16} seconds), while others had ridiculously short lifetimes, about 10^{-23} seconds, not much longer than the transit time of light across their diameters. As these new strongly interacting particles proliferated, only one tool could be brought to bear to try to make sense of them—symmetry.

TOO MANY FUNDAMENTAL PARTICLES

The first order of business in any science, such as zoology or botany or epidemiology, is to classify things. This means that you make lists of everything you have observed and then try to put these items into general related categories. For example, we might list animals according to whether or not they have backbones (vertebrates and invertebrates). Within this category we make a sub-list according to whether they have scales, feathers, fur, etc. Then we look for patterns among the lists. Eventually we discover relationships, and we can then formulate theories of their origins and try to explain the myriad patterns.

By the end of the 1950s there were three broad categories of "elementary" particles. First, there were a few *non–strongly interacting* matter particles (particles that don't participate in the strong interactions, that is, they do not interact with Yukawa's pions or any of their relatives). These were initially seen to be comparatively lighter-in-mass particles compared to the proton and neutron, so they were dubbed the "light ones," which in Greek is *leptons*. The class of leptons contained the electron, the muon, and two very hard-to-observe particles called the electron neutrino, ν_e, and the muon neutrino, ν_μ. Much later, in the mid 1970s, another pair joined and completed this class of leptons, called the "tau," τ, and the "tau" neutrino, ν_τ. Even though the tau is heavier, it shares the non–strongly interacting behavior of the electron and muon, and it fits into the lepton family.

By the 1970s accelerator experiments had confirmed that leptons were point-like, or structureless, objects down to the smallest accessible distance scales, about a hundredth of a millionth of a billionth of a centimeter (10^{-17} cm). In addition to the leptons there were two other particles that are strictly force carriers and that fall into a special class we call "gauge bosons." These include the well-known photon, the particle of light, and a hypothetical "graviton"—the particle of gravity.

The remaining particles comprising the vast list of strongly interacting particles were called the "hadrons" or "strong ones." All strongly interacting particles were found to have a finite nonzero "size" of about a hundredth of a thousandth of a billionth of a centimeter (0.2×10^{-13} cm). Various patterns began to emerge within that class of particles. Indeed, the patterns began to hint that hadrons are actually composed of smaller, more elementary objects deep down within another stratum that could not yet be resolved with the existing accelerator probes.

For a long time there was considerable resistance to the idea of any further substructure within the hadrons. No matter how high an energy probe struck the proton, it was not possible to "smash it into smithereens." All that happened was that other short-lived hadrons were produced in these collisions, and you ended up back with the original proton (or a neutron or pions) you started with. Evidently Democritus's idea of fundamental underlying "atoms" was breaking down with the discovery and properties of the hadrons. Very novel and Zen ideas emerged—perhaps hadrons are composed of *each other* in such a way that none is truly fundamental and yet all are? It was as if the world of hadrons were an Escher staircase, eternally going uphill, only to return again to the first step.

Connected to this idea was the notion that hadrons are not made of point-like objects but are more like the consistency of putty—deformable and malleable rather than point-like and hard. One of the most intriguing patterns among these objects could be explained if it was assumed that, as the putty rotates rapidly, it becomes drawn out into a kind of putty "string." Various quantum modes of motion of this "string" were studied, and it seemed to make sense—all of the hadrons could be explained as putty strings, and many of their properties were predicted and emergent from the idea. Thus was born, in attempting to explain hadrons, a new type of dynamical quantum theory, the string theory.

THE STRATUM OF QUARKS AND LEPTONS

But the long list of "too many strongly interacting particles" led some physicists, most notably Murray Gell-Mann and George Zweig of the California Institute of Technology, to assert that these were not fundamental. The long list of hadrons had certain patterns, like the recurring chemical properties of atoms, and hinted at the existence of yet another layer of the physics onion. Yet there was a serious problem with the idea of another stratum of nature—whatever comprised the strongly interacting particles could never be set free from the particles they composed by any experiment. Even the most powerful of particle accelerators, producing the most violent collisions, never liberated any of the hadronic innards, and instead simply produced more and more of the unstable hadrons.

Nonetheless, for a particular theoretical next layer of constituency of matter, whether real or purely mathematical, Gell-Mann introduced the term "quark."[1] In the early 1970s, through the theoretical insights of James Bjorken,[2] the first "photograph" of the inner world of the proton was taken at the Stanford Linear Accelerator by scattering very energetic electrons off of protons, a process known as "deep inelastic scattering." For the first time, the constituents of hadrons—the quarks—were seen. It was also observed that half of the constituents of the hadrons were something else—a mysterious electrically neutral component of these particles was detected. Could this be the "glue" that holds the quarks inside?

Initially, almost comically, the theoretical force carriers that bind quarks within hadrons were dubbed "gluons." Soon, however, by making a profound analogy with electric and magnetic forces generated by photons, a real theory of quarks and gluons, called "quantum chromodynamics" (QCD for short; QCD is a Yang–Mills gauge theory) took hold. Gluons joined the panoply of elementary particles and entered the list of bosons, like the photon and graviton. Gluons, indeed, generate the force that holds the quarks inside the strongly interacting particles.[3]

Quarks						Leptons		
q		mass	red	blue	yellow		mass	q
First Generation								
+2/3	up	2.3 MeV	u	u	u	• electron neutrino	< 2 eV	0
-1/3	down	4.8 MeV	d	d	d	• electron	0.511 MeV	−1
Second Generation								
+2/3	charm	1.27 GeV	c	c	c	• muon neutrino		0
−1/3	strange	95 MeV	s	s	s	• muon	0.105 GeV	−1
Third Generation								
+2/3	top	178 GeV	t	t	t	• tau neutrino		0
−1/3	beauty	4.18 GeV	b	b	b	• tau	1.78 GeV	−1

FIGURE A.35. Table of Quarks and Leptons. This exhibits the "generation structure" of the matter particles, by which a pair of "up"-type and "down"-type colored quarks fit together with a pair of "electron" and "neutrino"-type leptons.[4] In addition, there are the antiparticles, required by special relativity. Antiparticles have opposite electric charges and anti-colors, hence the blue quark has an antiparticle that is "anti-blue," which acts like a combination of red and yellow. The neutrinos have extremely tiny masses, expected to be less than about 2 electron volts.

As of today we have built many particle accelerators, some so powerful that we can clearly see the quarks and gluons deep inside the hadrons, like the nucleus inside the atom or the DNA inside of a living cell. The gluon force is not, however, like anything we have seen before. Unlike familiar electromagnetism, the gluonic force doesn't fall off like the inverse square law between two separated electric charges but is rather a constant force as we try to separate the quarks. This behavior ends up forbidding us from ever isolating the quarks. Quarks are confined forever inside of hadrons. In fact, the gluon force, when we rapidly rotate a hadron, becomes the putty-like string.

TODAY: THE PATTERNS OF QUARKS, LEPTONS, AND BOSONS

The elementary constituents of the hadrons are the quarks and gluons. Quarks and gluons are real, and their properties are measured, but they can never be set free from the prisons of the hadrons that they comprise. With quarks and gluons a more Democritus-styled explanation of the hadrons took hold, and this is the view that we have of them today. Quarks, like their sisters the leptons, are point-like and structureless matter particles.

We often refer to the quarks and leptons as "the matter particles." Each of these particles is a tiny gyroscope, each has spin 1/2 (see "Spin" below), in accordance with the rules of quantum mechanics. All the everyday matter in our world is essentially composed of the two quarks, the *up* and *down* (and *gluons*), and the one lepton, the *electron*. These quarks are distinguished by their electric charges and their masses. We always define the electron to have an electric charge of −1. In these units, the up quark (u) has an electric charge of +2/3, and the down quark (d) an electric charge of −1/3. The proton is therefore not an elementary particle but is rather a composite particle, built of three quarks in the pattern $u + u + d$ (or uud). Adding up the electric charges of the constituent quarks, we see that the proton charge is $+2/3 + 2/3 - 1/3 = +1$. Similarly, the neutron is composed of $u + d + d$, and the corresponding electric charge combination is $+2/3 - 1/3 - 1/3 = 0$.

Every particle in nature has a corresponding antiparticle. This was Dirac's famous discovery based upon unifying quantum theory with special relativity. The antielectron is the *positron* and has electric charge +1 and the same mass and spin as the electron. The antiquarks likewise have the opposite electric charges to their quark counterparts. We designate the anti-up quark as $\bar{u}$, and it has an electric charge –2/3, while the anti-down is $\bar{d}$, with electric charge +1/3.

The pions are composed of combinations of a quark and antiquark. We easily see that there are four possible quark-antiquark combinations involving u, d, $\bar{u}$, and $\bar{d}$, which are $\bar{u}d\,(-1)$, $\bar{u}u\,(0)$, $\bar{d}u\,(+1)$, $\bar{d}d\,(0)$. In quantum mechanics, neutral particle states often become "blended" (added together in particular ways), and the resulting composite particles are

$$\pi^+ \leftrightarrow \bar{d}u \quad \pi^0 \leftrightarrow \bar{u}u - \bar{d}d \quad \pi^- \leftrightarrow \bar{u}d \text{ and } \eta^0 \leftrightarrow \bar{u}u + \bar{d}d$$

The first three are the pions, and the fourth is called the "eta meson." All four are known well from experiment, and their quark composition accounts neatly for the pattern. In fact, from the masses of the pions and other mesons, we can deduce the masses of quarks themselves.

Only particular combinations of quark composites are observed experimentally to occur. In nature we only find objects containing three quarks (called *baryons*), or three antiquarks (called *anti-baryons*), or objects containing quark plus antiquark (called *mesons*). So the question arises: What is the nature of the strong force that holds the quarks together inside of the hadrons?

We find that each quark comes in "triplets." That is, there are three up quarks, three down quarks, three strange quarks, and so on. The additional label is the "color." Hence we say that there is a red up, a blue up, and a yellow up quark. This has nothing to do with visual colors of the rainbow but is a mnemonic description of the full symmetry of quarks.

The color of a quark is hard to detect, because any observed particle that the quarks compose, a hadron such as the proton and neutron and mesons, *always has a net color of zero*. For example, at any instant of time, the proton contains *uud*, but one quark is red, another blue, and another yellow, making an overall color-neutral state.

The antiquarks must be viewed as having anti-colors in the sense of the color wheel. So the anti–blue up quark is actually a red-yellow, or "cyan," object. Therefore, we can make color-balanced mesons by combining pairs of quark and antiquark. This simple rule explains the forms of the bound particles that we see. However, it also gives the clue to the fundamental theory of the strong interactions.

Gauge Bosons

q		mass		mass
Electroweak			**Strong (gluons)**	
0	photon	0 GeV	(red, anti-blue)	0 GeV
+1	W^+	80.4 GeV	(red, anti-yellow)	0 GeV
- 1	W^-	80.4 GeV	(blue, anti-red)	0 GeV
0	Z^0	90.1 GeV	(blue, anti-yellow)	0 GeV
			(yellow, anti-red)	0 GeV
Gravity			(yellow, anti-blue)	0 GeV
0	graviton	0 GeV	(red, anti-red) - (blue, anti-blue)	0 GeV
			(red, anti-red) + (blue, anti-blue) -2(yellow, anti-yellow)	0 GeV

Figure A.36. Table of Gauge Bosons. These are also the known "force carriers" and all are defined by "gauge" symmetries.

How do we establish that quark color exists if we can't see it? In fact, it was anticipated in the early days of the quark theory because of Pauli's exclusion principle. There exists a composite strongly interacting particle, whose properties Gell-Mann dramatically and precisely predicted in 1963. Experimentalists quickly confirmed the prediction at Brookhaven National Laboratory. This is the Ω^-, "omega-minus," and it contains three strange quarks, or *sss*. It is known that the quarks making up the Ω^- must move in a single common quantum state, or orbital, but without quark color this would be *strictly forbidden* by the Pauli exclusion principle. Yet the Ω^- does indeed exist. The only way out of this conundrum is the existence of quark color. If one s quark is red, the second blue, and the third yellow, making up the Ω^-, then there is no problem with the exclusion principle. There are many other ways in which the number of colors of quarks has been "counted" in experiment, and the result is always consistent with three.

We can think of a quark as though it lives in a "three-dimensional color space," whose three axes are labeled by the three colors. In this space, a quark can be thought of as an arrow (a vector) that can point in any color direction. If the quark is red, its arrow points along the red axis, if blue, then the blue axis, and so forth. However, in quantum theory the arrow can rotate and point in any direction. The color symmetry is just the collection of rotations that we can do to such a quark arrow (this is known as the "symmetry group SU(3)").

Now we generalize the subtle idea of "gauge symmetry," which governs electrodynamics. For the electron, the symmetry requires the introduction of the photon. The electron becomes a quantum blend with the photon. Gauge symmetry then implies that by "shaking" (accelerating) the electron, we can cause the emission of a gauge particle, or *gauge boson*, called the photon. This gives rise to all of the electromagnetic properties of matter.

But for quarks, we go further with this concept. We can rotate the quarks in color space, for example, we can rotate a pure red down quark into a blue down quark. We want this to be a "gauge symmetry," and this requires that we have additional particles that "undo" the changes we make on the red quark, keeping the overall result invariant. To have a color-gauge symmetry, we need 8 new gauge particles, called *gluons*.[5]

The physical gluons are emitted from the quarks when the quarks accelerate, like photons from electrons. But gluons carry off the old color of the quark and carry in a new color. So a gluon has (color + anti-color). So, when we turn a red quark into a blue quark, by emitting a gluon we simultaneously *create a (red, anti-blue) gluon*, so the net color is (red + anti-blue) + (blue) = (red), and therefore initial color of the quark is recovered.

Note that if a gluon collides with a quark, the gluon will be absorbed and the quark will be accelerated. It is, perhaps, one of the most astounding aspects of modern science that the simple idea of a symmetry yields up the photon, and quantum electrodynamics likewise yields the correct theory of the strong interactions when it is adapted to quark color. This theory is called quantum chromodynamics (QCD), and it is a stunning success.

Quarks thus interact with one another by the exchange of gluons. We can draw the appropriate Feynman diagrams and learn how to compute them. The force is strong because the "color charge," g, the analog of the electric charge, is large.

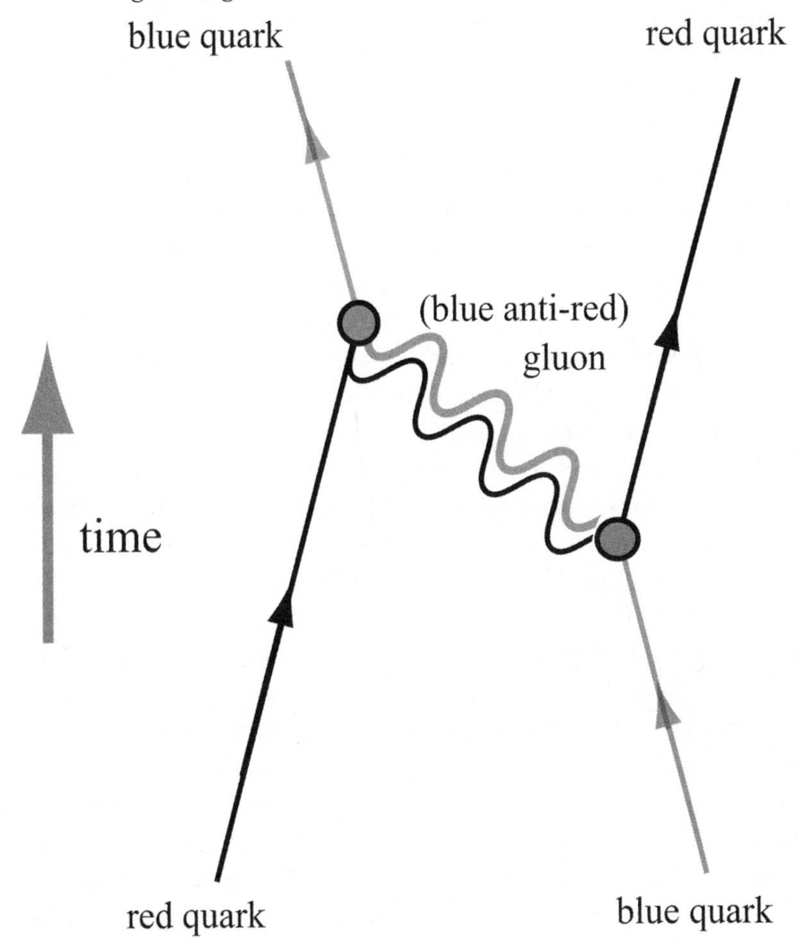

Figure A.37. Quark Exchanging Gluons. A red quark scatters off of a blue quark. The quarks exchange color through the (red, anti-blue) gluon that hops between them, giving rise to the strong force that confines the quarks. The force becomes weaker at higher energies, a prediction of the theory that is well confirmed by experiment. The proton is held together by gluon exchange among the quarks. Every 10^{-24} seconds or so, a gluon hops between quarks in the proton.

One of the most remarkable discoveries about QCD is that the coupling strength of the interaction of the quarks to the gluons, denoted by g, becomes weaker as we push the quarks together to extremely short distances. This is known as "asymptotic freedom." Conversely, at large distances, the quark-gluon coupling becomes very large. This leads to a strong pull on the quarks, preventing them from being separated and isolated in the lab. It also turns out that, because of this strong coupling, only quantum-bound states that are composed of quarks that have an exact color neutrality—a perfect balance of all three quark colors at any instant in time—can exist. This means we can have only combinations of (*rby*), which are the *baryons*, or $(\overline{r}\overline{b}\overline{y})$, which are the *anti-baryons*; by $(\overline{q})$ we mean the anti-color of (q), or the color-neutral quantum combination of $\pi^+ \leftrightarrow \overline{d}u$, which are the mesons. The color gauge theory, QCD, therefore neatly explains the pattern of strongly interacting particles found in nature, and why the quarks can never be really freed from their prisons.

While it is very hard to compute the properties of the theory when g is large, the fact that g becomes small at short distances means that fairly precise calculations using Feynman diagrams can be performed revealing the collisions and scattering of individual quarks at very high energies. It also means that at very high energies, for example, in the collisions produced at the LHC, the individual quarks and gluons collide and leave traces of their collisions. This leads to a bizarre phenomenon, nature's own version of a prison break, known as a *quark jet* (*gluon jets* can also occur).

At the LHC, a proton with seven trillion electron volts of energy (7 TeV) collides head-on with a proton of the same energy. At the highest energies, or shortest instants of time, the individual quarks are resolved and behave as though they were almost free particles. Therefore, collisions occur in which a pair of quarks, perhaps a u and a d collide head-on. This quark and antiquark are scattered through very large angles, ripped out of the proton and antiproton, while the remaining debris, the other quarks and gluons of the original proton and antiproton, continue to move forward in their original directions of motion. For a brief moment the quarks are free, moving at very high energies, and they can travel perhaps a hundred times the distance within which they are normally confined, away from the debris of quarks and gluons of the shattered proton and antiproton. The quarks, for a brief moment in time, have broken out of their confining prison cells.

But then the interaction becomes strong and the vacuum itself begins to rip apart in the vicinity of the collision. Pairs of quarks and antiquarks and gluons are ripped out of the vacuum, a turbulent plasma of matter from the point of collision, like the long arm of the law apprehending the escapees. The liberated quarks become shackled by this flurry of new matter and antimatter. Soon all quarks and gluons are recaptured, reassigned to new pions and protons and neutrons. The liberation of the quarks is over.

Nonetheless, the indelible footprint of the escaped quarks remains. Two very well-defined blasts of particles, called *jets*, mostly composed of pions, stream off into space in the directions of the original u and d escapees. These jets of particles clearly mark the original quark paths and carry the full energy of the temporarily liberated quarks. These jets are the conspicuous tracers of high-energy quark, and gluon, collisions.

At the LHC a pair of gluons can collide to produce a Higgs boson. The decay signature of the Higgs boson is reconstructed in the detector. There are many decay modes of the Higgs boson, but the one first glimpsed at the LHC was a pair of gammas, two high-energy photons, in the process $g + g \rightarrow \text{higgs} \rightarrow \gamma + \gamma$. In this way, nature's most enigmatic particle, the Higgs boson, is pulled from the vacuum's Higgs field that surrounds us and gives mass to all the other particles.

SPIN

Any rotating body has spin—a top, a CD player, the earth, the washing machine basin on the rinse cycle, a star, a black hole, a galaxy—all have spin. So, too, quantum particles, molecules, atoms, nuclei of atoms, the protons and neutrons in the nucleus, the particle of light (photons), electrons, the particles inside of protons and neutrons (quarks, gluons), etc. But while large classical objects can have any amount of spin, and can stop spinning altogether, quantum objects have "intrinsic spin" and are always spinning with the same total intrinsic spin.

An elementary particle's spin is one of its defining properties. We can never halt an electron from spinning, else it would no longer be an electron. However, we can rotate a particle in space, and the value of its spin, as projected along any given axis in space, will change, just as it does for a classical spinning top. The difference in quantum physics is that we can only ask what value does the spin have when projected along a given axis, because that is what we can measure—asking about things we cannot measure is meaningless in quantum physics.[6]

Let's discuss the rotational motion of a classical object. Linear physical motion is measured by something called *momentum*. In Newtonian physics this is simply mass times velocity. Note that this combines the concept of matter (mass) and concept of motion (velocity), so it represents a kind of measure of "physical motion." This is a vector quantity, since the velocity is a vector, having both a magnitude (speed) and a direction (of motion) in space. In general a vector can be visualized as an arrow in space with both magnitude and direction.

Likewise, physical rotational motion is measured by a (pseudo-vector) quantity called "angular momentum." Classically, angular momentum involves the way in which mass is distributed throughout the object, which is the "moment of inertia." If a body is large, with a large radius, when it spins there is a lot more matter spinning than if the same amount of mass were distributed within a smaller radius. So, not surprisingly, the moment of inertia, I, increases with the size of the body. In fact, it's mass times "the (approximate) radius of the body squared," or roughly $I = MR^2$ with M the mass and R the "radius" of the body. This can be made very precise using calculus.

Spin also involves the "angular velocity," how fast the object is actually rotating. Angular velocity is usually denoted by ω (omega) and is "so and so many radians of rotation per second." (360 degrees equals 2π *radians*. So, for example, 90 degrees corresponds to $\pi/2$ radians; radians are a more mathematically natural way to measure angles than degrees because a circle with a radius of one has a circumference of 2π. Therefore, spin is just the product of the moment of inertia times the angular velocity, or $S = I\omega$. (Compare: momentum is mass times velocity and describes motion in a straight line, while spin is moment of inertia times angular velocity—these are very similar constructs.) Spin is also a vector quantity, pointing along the axis of the spin rotation. Here we use the "right-hand rule" to establish the direction of the spin vector: curl the fingers of your right hand in the direction of the spinning motion and your thumb will point in the direction of the spin vector.

Spin is a form of angular momentum, which is a *conserved quantity* (like energy and momentum) such that the total angular momentum of an undisturbed isolated system remains forever constant. As a consequence of this, we see that an ice skater, viewed as a physical system, can dramatically increase her spin motion (angular velocity) as she draws her arms inward. The spin angular momentum is $S = I\omega = MR^2\omega$, which must stay the same as she pulls her arms in. Pulling her arms in decreases R, while M stays the same. So, the angular velocity ω must increase to compensate the decrease as the rotational velocity increases. In fact, R^2 becomes four times smaller if the skater simply decreases her arms' outward distance, R, by a half, so her angular velocity must increase approximately fourfold, which is why this is such a dramatic stunt.

Angular momentum, which was a continuously varying quantity in Newtonian physics, changes its character drastically in quantum mechanics—it becomes quantized. *Angular momentum is always quantized in quantum mechanics.* All observed angular momenta as measured along any *spin axis* are discrete multiples of $\hbar = h/2\pi$, where h is Planck's constant. All the particle spin and orbital states of motion we find in nature have angular momenta that can have only the exact values

$$0,\ \frac{\hbar}{2},\ \hbar,\ \frac{3\hbar}{2},\ 2\hbar,\ \frac{5\hbar}{2},\ 3\hbar, \ldots \text{ and so forth.}$$

Angular momentum is always either an *integer* or a *half-integer* multiple of $\hbar$ in nature. We don't see this quantization effect for very large classical objects because they can have such enormous angular momenta, many times greater than $\hbar$. Only at the level of exceedingly tiny systems, atoms, or the elementary particles themselves, do we observe the quantization of angular momentum.

Angular momentum is therefore an intrinsic property of an elementary particle or an atom. All elementary particles have spin angular momentum. We can never slow down an electron's rotation and make it stop spinning. An electron always has a definite value of its spin angular momentum, and that turns out to be, in magnitude, exactly $\hbar/2$. We can flip an electron and then find its angular momentum is pointing in the opposite direction, or $-\hbar/2$. These are the only two observable values of the electron's spin when measured along any chosen direction in space. We say that "the electron is a spin-1/2 particle," because its angular momentum is the particular quantity, $\hbar/2$.

Particles that have *half-integer multiples* of $\hbar$ for their angular momentum, that is,

$$\frac{\hbar}{2}, \frac{3\hbar}{2}, \frac{5\hbar}{2}, \ldots \text{ and so on,}$$

are called *fermions*, after Enrico Fermi, who helped pioneer these concepts (with Pauli and Dirac). The main fermions we encounter in most of our discussions are the electron, the proton, or the neutron (and quarks, which make up the proton and the neutron, etc.), and each has angular momentum $\hbar/2$. We refer to all of these as "spin-1/2 fermions."

Particles, on the other hand, that have angular momenta that are *integer multiples* of $\hbar$, such as $0, \hbar, 2\hbar, 3,\hbar, \ldots$ and so on, are called *bosons*, after the famous Indian physicist Satyendra Nath Bose, who was a friend of Einstein and who developed some of these ideas. There is a profound difference between fermions and bosons that we'll encounter momentarily. Typically, the only particles that are bosons and that will concern us presently are particles like the photon, which has "spin 1," or one unit of $\hbar$ angular momentum; the quantum particle of gravity, the graviton, which has yet to be detected in the lab, and has "spin 2," or $2\,\hbar$ units of angular momentum; and other particles that are made of quarks and antiquarks, called mesons, that have "spin 0," or 0 units of angular momentum. Orbital motion also has angular momentum. All orbital motion, in quantum theory, has integer units of $\hbar$ for angular momentum, hence, $0, \hbar, 2\hbar, 3,\hbar, \ldots$ and so on.

EXCHANGE SYMMETRY

Elementary particles are so fundamental that they have no identifying labels. For example, any two electrons cannot in principle be distinguished from each other. There is no difference between any two electrons in the universe. The same is true of photons, muons, neutrinos, quarks, etc. The quantum effect of this identity symmetry depends strongly upon spin.

Now, in everyday life, the category of "things" that we encounter called "dogs" is very large, and no two dogs are identical. However, all electrons are precisely identical to each other. Electrons carry only a very limited amount of information. Any given electron is *exactly* identical to any other electron. The same is true of the other elementary particles. Therefore, any physical system must be symmetrical, or invariant, under the swapping of one such particle with another. In a sense, nature is very simple-minded in the way it treats electrons in that it doesn't know the difference between any two (or more) electrons in the whole universe.

"Exchange symmetry" implies that swapping two identical particles *must leave the laws of physics invariant because the particles are identical.* At the quantum level this implies that our swapped particle waves must give the same observable probability as the original. But probability involves taking the "square" of the waves, or more properly, the square of their "wave functions."[7] This condition, however, implies two possible solutions for the effect of the exchange on the wave function, that is, the exchanged wave can either be *symmetrical*, +1 times the original one, or else it can be *anti-symmetrical*, –1 times the original one. Either case is allowed, in principle, because we can measure only the probabilities (the squares of wave functions). Quantum mechanics allows both possibilities, so nature finds a way to offer both possibilities, and the result is astonishing.

BOSONS

For *bosons*, upon swapping two particles in the wave function, we would get the + sign.[8] With this result, we find an important effect—two identical bosons can be located in the same quantum state. In fact, by considering lots of bosons localized in the same region of space, described by one big wave function, we can actually prove that the *most probable place for all the bosons in a system is piled on top of one another*. So, it is possible to coax a lot of identical bosons to share the same little region in space, almost an exact pinpoint in space. Or, the identical bosons can be coaxed readily into a quantum state with the exact same value of momentum. Thus, we say that bosons *condense* into compact, or "coherent," states. This is called *Bose–Einstein condensation*.

There are many variations on Bose–Einstein condensation and all kinds of phenomena that have in common many bosons in one quantum state of motion. Lasers produce coherent states of many, many photons all piled into the same state of momentum, moving together in exactly the same state of momentum at the same time. Superconductors involve pairs of electrons bound by crystal vibrations (quantum sound) into spin = 0 bosonic particles (called "Cooper pairs"). In a superconductor the electric current involves a coherent motion of many of these bound pairs of electrons sharing exactly the same state of momentum. Superfluids are quantum states of extremely low-temperature bosons (as in liquid ^{4}He), in which the entire liquid condenses into a common state of motion that becomes completely frictionless. It has to be the isotope ^{4}He in order to get a superfluid (2 protons + 2 neutrons in the nucleus), because the isotope ^{4}He is a boson, while the other common isotope ^{3}He is not (with 2 protons + 1 neutron in the nucleus, it is a fermion; see below). Bose–Einstein condensates can occur in which many bosonic atoms condense into ultra-compact droplets of very large density, with the particles piling on top of one another in space.

FERMIONS

For *fermions* the rule is that we get the (–) sign in front of the wave function. This holds for any particle with fractional spin, such as the electron with spin 1/2. From this we can prove that no two identical fermions can occupy the same quantum state at the same time. This is known as the Pauli exclusion principle, after the brilliant Austrian-Swiss theorist Wolfgang Pauli. Pauli proved that his exclusion principle for spin 1/2 comes from the basic rotational symmetries of the laws of physics. It involves the mathematical details of what spin-1/2 particles do when they are rotated. Swapping two identical particles in a quantum state is identical to rotating the system by 180^0 in certain configurations, and the behavior of the spin-1/2 wave function then gives the minus sign (see note 8).

The exclusion property of fermions largely accounts for the stability of matter. For spin-1/2 particles there are two allowed states of spin, which we call "up" and "down" ("up" and "down" refer to any arbitrary direction in space). Thus, in an atom of helium, we can get two electrons into the same lowest-energy orbital state of motion. To get the two electrons in one orbital requires that one electron has its spin pointing "up," and the other has spin pointing "down." However, we *cannot then insert a third electron* into that same orbital state because its spin would be the same, either up or down, as one of the two electrons already present. The exchange symmetry minus sign would force the wave function to be zero.

In other words, if we try to exchange the two electrons whose spins are the same, the wave function would have to equal minus itself and must therefore be zero! Hence, for the next atom, lithium, the third electron must go into a new state of motion, that is, a new orbital. Thus, lithium has a *closed inner orbital, or "closed shell"* (i.e., a helium state inside of it), and a sole outer electron. This outer electron behaves much like the sole electron in hydrogen. Therefore, *lithium and hydrogen have similar chemical properties*. We thus see the emergence of the Periodic Table of Elements. If electrons were not fermions and did not behave this way, every electron in the atom would rapidly collapse into the ground state. All atoms would behave like hydrogen gas. The delicate chemistry of organic (carbon-containing) molecules would never happen.

Yet another extreme example of fermionic behavior is that of the neutron star. A neutron star is formed as the core of a giant supernova implodes while the rest of the star is blown out into space. The neutron star is made entirely of gravitationally bound neutrons. Neutrons are fermions, with spin 1/2, and again the exclusion principle applies. The state of the star is supported against gravitational collapse by the fact that it is impossible to get more than two neutrons (each with spins counter-aligned) into the same state of motion. If we try to compress the star, the neutrons begin to increase their energies because they cannot condense into a common lower-energy state. Hence, there is a kind of pressure, or resistance, to collapse, driven by the fact that fermions are not allowed into the same quantum state.

GAUGE SYMMETRY

It has been known for several hundred years that *electric charge is conserved* in any physical process. This conservation law is fundamental to the classical theory of electric and magnetic fields, or electromagnetism.

We see an example of electric charge conservation when we consider the decay of the neutron, $n^0 \rightarrow p^+ + e^- + \overline{\nu}^0$. The neutron is electrically neutral, having zero electric charge. When it decays, we are left with a positively charged proton, a negatively charged electron, and a neutral (anti) neutrino. The positive charge of the proton identically equals the opposite of the negative electron charge, and the neutrino has zero electric charge, so the final products of the neutron decay have a zero total electric charge. Electric charge conservation is an *exact conservation law in all physical processes*—we have never seen a net gain or loss of electric charge in any physical process. The existence of this conservation law implies some hidden symmetry in nature.

Electromagnetism, or "electrodynamics," is the physical description of electric and magnetic fields, and electric charges and currents, and it was formulated in a classical (non-quantum) framework over the entirety of the nineteenth century. The pinnacle achievement is usually considered to be the formulation Maxwell's equations, discovered in 1861 by James Clerk Maxwell, a succinct and complete set of equations that summarized all known aspects of electrodynamics, which allow us to compute the electric and magnetic fields anywhere in space and time, given any choice of electric charge and electric current distributions.[9]

Maxwell's classical theory of electrodynamics makes no sense without the conservation law of electric charge. The underlying continuous symmetry that leads to this, however, appeared, at first, to be somewhat obscure. Electric charges are the sources of *electric fields*, much like mass is the source of a gravitational field in Newton's theory of gravity. An *electric field* is just the electric force exerted on an electric charge at any point in space. When electric charges move, they become electric currents and produce *magnetic fields*. Magnetic fields, in turn, produce forces on moving electrons (electric currents). In fact, a pure electric field in space becomes a combined electric and magnetic field if we simply move through it.

The Maxwell theory does not allow solutions to its equations in which a source or a sink, an electric charge, simply disappears into nothingness. Even if an electric charge falls into a black hole, the black hole itself will have the same value of the electric charge that it swallowed.

If we probe deeper into the structure of Maxwell's theory, however, we find that there is something even more fundamental than the electric and magnetic fields called a *gauge field.* The gauge field is related to the electric and magnetic fields in a peculiar way: If we are given the gauge field in any region of space and time, we can always calculate the values of the electric and magnetic fields in that region. However, we cannot reverse this process. That is, given electric and magnetic fields in the same region of space and time, we cannot determine exactly what gauge field produces them. In fact, we can always find *an infinite number of gauge fields* that would produce the same observed electric and magnetic fields.

Moreover, while electric and magnetic fields are easily measured in the lab, we cannot directly measure the gauge field by theory or experiment. Even a zero value everywhere for the electric and magnetic fields, that is, a vacuum, does not determine the value of the gauge field—infinitely many different gauge fields exist that produce zero values of the electric and magnetic fields. The gauge field is therefore a hidden field, not amenable to any measurement that would determine its exact form.

The concept of a gauge field was first considered as a tool for conveniently expressing electric and magnetic forces by various scientists in the early to mid-1800s. Often different people would write down different gauge fields, in different forms, and it was always unclear whether or not they were describing different phenomena. In 1870, Hermann Ludwig Ferdinand von Helmholtz, a famous contributor to the theory of electromagnetism, showed that different forms of gauge fields can lead to the same physical consequences, that is, to the same electric and magnetic fields. One can *continuously transform one gauge field into another, and the physics stays the same.* This is essentially the first example of a new symmetry transformation of electrodynamics—a "gauge transformation"—though its implication as a fundamental symmetry of nature was not appreciated at the time.[10]

In fact, if we turn this around and insist that, as *a symmetry principle*, the gauge field must always be a hidden field and can never be determined unambiguously, then we do find something remarkable: it is this gauge symmetry that implies that electric charge must be conserved! We can continuously *transform* our chosen gauge field into another one, without changing the values of the electric and magnetic fields, and this is the symmetry that leads to the conservation of electric charge. This hidden symmetry is called "local gauge invariance."[11]

It was in the twentieth century, with the development of quantum mechanics, and the effort to include both the electron and electromagnetism into one completely consistent theory, that the symmetry of gauge invariance emerged as the overarching theme. In fact, this has been the dominant theme in all of twentieth-century physics—*all forces* are now known to be governed by "gauge symmetries" and are called "gauge theories."

All particles are described in quantum theory by waves, through their *wave functions*. The wave function is denoted by $\psi(x, t)$ and *is a complex number–valued function of space and time*. The probability of finding the particle at space location x and at time t is determined by the mathematical square of the wave function $|\Psi(x, t)|^2$. The information about the particle's momentum is determined by the *wavelength* of the wave, and the energy by the frequency through the formulae $E = hf$, "energy equals Planck's constant times the frequency," and $p = h/\lambda$, "momentum equals Planck's constant divided by the wavelength." Despite the fact that this energy and momentum information is always present in the wave function and can readily be extracted from it by differentiating it with respect to t or x, we can never measure the wave function directly because the wave function involves complex numbers that don't make sense as physical observables. Only the (absolute squared) magnitude of the wave function, which is the *probability*, can actually be measured.

We can ask: "What would happen if we somehow changed the phase of the wave function *without changing the observable probability at any point in space and time*?" We keep the probability of finding the electron at any point in space the same. We call this a "gauge transformation." But, in making this change, there is apparently nothing invariant here. This would affect the derivatives of the wave function with respect to t and x, and those determine the energy and momentum. This is evidently not a symmetry of the original quantum state, but rather it seems to produce a new quantum state with different observable energy and momentum.

Let us now suppose that there is *some other quantum particle wave* that modifies the derivatives with respect to t and x. And let us further suppose that when we change the electron's wavelength or frequency, *we are simultaneously required to modify the new field in such a way as to keep the derivatives with respect to t and x the same*. The net effect is that we have kept the probability, the energy and the momentum invariant under our transformation. Together with the gauge particle, we can maintain both the original incoming total energy and the momentum, even though we scramble the unobservable phase of our electron's wave function. Thus, the term "gauge" means that the actual determination of the physical momentum of the electron requires the presence of the calibrating "gauge" field. Only the electron wave function, together with the "gauge" field, yields a physically meaningful description of the electron. The presence of the new gauge field in the derivatives causes the interaction of the photon with the electron (see note 11).

The gauge theory asserts that, if the electron is given a physical kick, if an electron is *accelerated*, then the gauge field is actually shaken off—it is emitted as an independent particle wave with a physical momentum of its own, and the electron recoils to conserve energy and momentum. The gauge field becomes a true physical entity and is radiated out into space. From the point of view of a distant observer, an accelerated electron has radiated a new particle, the *photon*.

Light is emitted from accelerated charges. This occurs in countless physical processes, such as the *scattering* of an electron off of an atomic nucleus, or an atom, or another electron. It can be observed readily in the laboratory. At very low energies, it is the way in which electrons emit the photons from a campfire. Accelerated electrons radiate the microwaves that heat our coffee in a microwave oven, or transmit the evening news into our living rooms, or cause the sun to shine.

We can graphically represent a physical process by a set of Feynman diagrams that represents the quantum computation. These diagrams tell us precisely how to compute the quantum outcome, the probability of a given process, provided that the strengths of the interactions are known and are not too large. We can often visualize a process through Feynman diagrams even when we cannot compute the result. A graduate student, writing from Cornell University where Feynman developed this technique, commented, "At Cornell, even the janitors use Feynman diagrams."[12] With the full machinery of Feynman diagrams we can compute the scattering rate for two beams of electrons to arbitrary precision, including many diagrams that represent detailed quantum corrections to the basic result. The experimentalist can compare the calculations with the results measured in the lab, and these are found to agree to extremely high precision.

YANG-MILLS GAUGE THEORY

The modern era of gauge theories began with a remarkable paper of Chen Ning Yang and Robert Mills in 1954.[13] These authors asked a straightforward question: "What happens if we extend the gauge symmetry of the electron to larger symmetries?" The symmetry of electrodynamics involves, as we have seen, the phase of the electron wave function. This is called "U(1) symmetry."

Yang and Mills turned to the next more complicated symmetry, "SU(2)," the symmetry of the rotations sphere in three real dimensions (or the symmetry of the rotations of two particles, such as (u, d) quarks or (ν_e , e^-), that is, rotations in 2 complex dimensions). It turns out that this symmetry leads to a more general form of a quantum gauge theory called a "Yang–Mills theory." SU(2) has three gauge fields, hence three photon-like objects, and now the gauge fields themselves carry charges, unlike the case of electrodynamics in which the photon carries no electric charge. Moreover, the Yang–Mills construction works for any symmetry. Symmetry thus becomes partially fundamental to the basic structure of a quantum theory of forces.

In the Standard Model electroweak theory, the symmetry is the "product group" of SU(2) × U(1), with 4 gauge fields, W^+, W^-, Z^0, and γ fully and accurately described by the Yang–Mills theory. The Higgs boson, as we have seen, causes the W^+, W^-, and Z^0 to become heavy, while the photon γ remains massless. Likewise, as we've seen, the quarks carry 3 "colors," and the resulting SU(3) gauge theory has 8 gauge bosons, known as the gluons.

Indeed, all known forces are based upon gauge theories. Yet, there are four completely different structures, or *styles*, of gauge invariance. Einstein's theory of gravity contains a coordinate system invariance, that is, it doesn't matter what coordinate system you use, or how you choose to move, inertially or non-inertially through space and time, to describe nature. This leads to gravity as a bending and reshaping of geometry, governed by the presence of energy (equivalent to mass) and matter. Particles must then emit and absorb *gravitons*, which are the gauge fields, or the "quanta," of gravity. The Newtonian gravitational theory is recovered only as an approximation at low energies (slow systems, without too much mass). The description of the remaining nongravitational forces in nature is based upon the Yang–Mills theory of SU(3) × SU(2) × U(1) as codified by the Standard Model.

THE WEAK FORCE AS A GAUGE THEORY

Let's now consider in a little more detail how the *weak interactions* are described by a gauge symmetry that unifies them together with the electromagnetic force. Taken together, the quarks, leptons, and the gauge symmetries (including Einstein's general relativity) provide a complete accounting of all observed laboratory physics to date and define what is called the "Standard Model."[14]

Recall that, subsequent to Becquerel et al., yet more than 65 years ago, Enrico Fermi wrote down the first descriptive quantum theory of the "weak interactions." Fermi had to introduce a new fundamental constant into physics to specify the overall strength of the weak interactions, called G_F, and it represents a fundamental unit of mass, which sets the scale of the weak forces, about 175 GeV.

In the 1960s the weak forces were found to involve a gauge symmetry based upon the symmetries SU(2) × U(1) (by Sheldon Glashow, Abdus Salam, and Steven Weinberg, and this was perfected as a quantum theory by Gerhard 't Hooft and Martinus Veltman). Let us now describe the gauge symmetry of the weak interactions.

We see that, within each generation, the quarks and leptons are paired. That is, the red up quark is paired with the red down quark, the electron neutrino with the electron, the charm quark with the strange, the top with the bottom, and so forth. We thus imagine that the electron and its neutrino are a single entity that lives in a two-dimensional space, with one axis meaning "electron" and the other "electron neutrino." The quantum state is an arrow in this space that can point in any direction. When the arrow points along the electron axis, we have an electron. Rotating the arrow, we have a neutrino. The rotations we can do on the arrow form the symmetry group, called SU(2).

So we now imagine an electron neutrino particle wave with a given momentum and energy. Then we perform a *gauge transformation* that rotates this into an electron, which has negative charge, and also scrambles the electron momentum and energy. To make this into a symmetry, we need to introduce a gauge field, the W^+ that can restore the total energy and momentum, and rotate the quantum arrow back to its original electrically neutral "electron neutrino" direction. In a sense, the gauge field rotates the coordinate axes, so the arrow is now pointing back in the original direction, relative to the coordinate system, and we get back the original neutrino we started with. This is completely analogous to what we do with quark color, where the gauge rotation from one color to another is compensated by the gluon field. This requires a total of three new gauge fields, W^+, W^-, Z^0, in addition to the photon γ. In fact, electrodynamics and the weak interactions now become blended together into one combined entity called the "electroweak interactions."

There is, however, an enormous difference between the photon and these three new gauge fields. The photon is a massless particle, while the W^+, W^-, Z^0 are very heavy particles. The forces that are produced by the quantum exchange of W particles between quarks and leptons give rise to the weak force that Fermi was describing 65 years ago. As we've seen, the Higgs field in the vacuum causes the W^+, W^-, Z^0 to become massive.

The strength of the Higgs field in the vacuum is already determined by Fermi's theory to be 175 GeV. The field implies the existence of a new particle, the Higgs boson, the necessary quantum of the Higgs field. All the matter particles, and the W^+, W^-, Z^0, get their masses by interacting with the vacuum-filling Higgs field (unlike a superconductor, however, the photon does not interact with this particular field and remains massless). The Higgs field is "felt" by the various particles through their "coupling strengths." For example, the electron has a coupling strength with the Higgs field, *ge*. Therefore, the electron mass is determined to be $me = ge \times$ (175 GeV). Since we know $me = 0.0005$ GeV, we see that $ge = 0.0005/175 = 0.0000029$. This is an extremely feeble coupling strength, so the electron is a very-low mass particle. Other particles, like the top quark, which has a mass $mtop = 175$ GeV, has a coupling strength almost identically equal to one (suggesting that the top quark is playing a special role in the dynamics of the Higgs field). Still other particles, like neutrinos, have nearly zero masses and therefore nearly zero coupling strengths.

All of this sounds like a spectacular success, and it is, but there is a slight shortcoming—there is, at present, *no theory for the origin of the coupling constants*, such as that of the electron g_e. These appear only as input parameters in the Standard Model. We learn almost nothing about the electron mass, swapping the known experimental value, 0.511 MeV, for the new number, $ge = 0.0000029$. Furthermore, we are clueless as to what generates the mass of the Higgs boson itself.

The Standard Model did successfully predict the coupling strength of the W^+, W^-, Z^0 particles to the Higgs field. These coupling strengths are determined from the known value of the electric charge and another quantity, called the weak mixing angle, measured in neutrino scattering experiments. So the masses of W and Z, M_W and M_Z (note that the W^+ and W^- are particle and antiparticle of each other and must have the same identical masses; the Z^0 is its own antiparticle), are predicted (correctly) by the theory. The W^+ and W^- have a mass of about 80 GeV, and the Z^0 has a mass of about 90 GeV. These have been measured to very high precision in experiments at CERN, SLAC, and Fermilab.

Symmetry and its spontaneous breaking through of the Higgs particle, therefore, completely controls the mass generation of all the particles in the universe. And it appears that it is the Higgs boson, the quantum of the Higgs field, that was discovered in the two experiments, ATLAS and CMS, at CERN on July 4, 2012, with a mass of m_h = 126 GeV. "What generates the Higgs boson mass?" is now the most important scientific question of our time.

NOTES

Note: We view the chapter notes as a study guide to the subject of particle physics and related sciences. Particle physics was the origin of the World Wide Web, and today the Web, and also *Wikipedia*, have become the most readily accessible, up-to-date, and reasonably reliable sources of information about the sciences. We have therefore provided many references, without apology, to related *Wikipedia* articles, and we urge the reader to seek out additional sources that are referred to within these articles.

CHAPTER 1. INTRODUCTION

1. We recommend an excellent account of the remarkable story of modern economics: David Warsh, *Knowledge and the Wealth of Nations: A Story of Economic Discovery* (W. W. Norton & Company, 2007). We thank our distinguished colleague, Dr. Richard Vidal, for bringing this to our attention.

2. See "Tim Berners-Lee," http://en.wikipedia.org/wiki/Tim_Berners-Lee; Berners-Lee and his associates, who helped design the Internet protocols and early browser, Robert Kahn, Vinton Cerf, Marc Andreessen, and Louis Pouzin, recently received the inaugural Queen Elizabeth Prize for Engineering ($1.5 million each), http://blogs.wsj.com/tech-europe/2013/03/18/berners-lee-wins-1-million-engineering-prize/. See also "Mosaic" (web browser), http://en.wikipedia.org/wiki/Mosaic_browser (sites last visited 3/20/2013).

3. Al Gore's contribution was to create and pass legislation called the "High-Performance Computing and Information Act of 1991" (also called the "Gore Act") to free up and use government-funded networks of computers, known as ARPANET, to the people. The World Wide Web really began as a combination of Berners-Lee's software and the ARPANET hardware. The Gore contribution to the Internet can be found on the Internet: http://en.wikipedia.org/wiki/Al_Gore_and_information_technology; see also "ARPANET," http://en.wikipedia.org/wiki/ARPANET (sites last visited 3/1/2013).

4. Here are some references on the SSC and its demise: "Lots of Reasons, but Few Lessons," http://www.sciencemag.org/content/302/5642/38.full; see "SSC," http://

en.wikipedia.org/wiki/Superconducting_Super_Collider; "How Close Was the Vote to Cancel the Superconducting Super Collider?" http://www.quora.com/How-close-was-the-vote-to-cancel-the-Superconducting-Super-Collider (sites last visited 3/8/2013). Quoting from the latter source,

> The House of Representatives voted three times in 1992 and 1993 to kill the SSC; the final pivotal vote was 159–264 (139 Cong. Rec. H8124 (daily ed. Oct. 19, 1993)). The Senate voted to rescue it each time; their last vote in favor was 57–42 (139 Cong. Rec. S12,760 (daily ed. Sept. 30, 1993)). In 1993, the two houses met in a conference committee twice; the first time the Senate negotiators won and the SSC was left in the bill. The second time the House won. In the end the conference report was adopted by both houses with large majorities: 332–81 in the House, 139 Congressional Record H8435 (daily ed. Oct. 26, 1993), and 89–11 in the Senate, 139 Congressional Record S14483 (daily ed. Oct. 27, 1993).

5. See the historical website of CERN: http://public.web.cern.ch/public/en/about/History-en.html (site last visited 3/1/2013). Quoting from this source:

> CERN is run by 20 European Member States, but many non-European countries are also involved in different ways. Scientists come from around the world to use CERN's facilities. The current Member States are: Austria, Belgium, Bulgaria, the Czech Republic, Denmark, Finland, France, Germany, Greece, Hungary, Italy, the Netherlands, Norway, Poland, Portugal, the Slovak Republic, Spain, Sweden, Switzerland and the United Kingdom. Romania, Israel and Serbia are candidates to become Member States of CERN. Member States have special duties and privileges. They make a contribution to the capital and operating costs of CERN's programmes, and are represented in the Council, responsible for all important decisions about the Organization and its activities.
>
> Some states (or international organizations) for which membership is either not possible or not yet feasible are Observers. "Observer" status allows non-Member States to attend Council meetings and to receive Council documents, without taking part in the decision-making procedures of the Organization. Scientists from institutes and universities around the world use CERN's facilities. Physicists and their funding agencies from both Member and non-Member States are responsible for the financing, construction and operation of the experiments on which they collaborate. CERN spends much of its budget on building new machines (such as the Large Hadron Collider), and it only partially contributes to the cost of the experiments. Observer States and Organizations currently involved in CERN programs are: the European Commission, India, Japan, the

Russian Federation, Turkey, UNESCO and the USA. Non-Member States with co-operation agreements with CERN are: Algeria, Argentina, Armenia, Australia, Azerbaijan, Belarus, Bolivia, Brazil, Canada, Chile, China, Colombia, Croatia, Cyprus, Ecuador, Egypt, Estonia, Former Yugoslav Republic of Macedonia (FYROM), Georgia, Iceland, Iran, Jordan, Korea, Lithuania, Malta, Mexico, Montenegro, Morocco, New Zealand, Pakistan, Peru, Saudi Arabia, Slovenia, South Africa, Ukraine, United Arab Emirates and Vietnam. CERN also has scientific contacts with: China (Taipei), Cuba, Ghana, Ireland, Latvia, Lebanon, Madagascar, Malaysia, Mozambique, Palestinian Authority, Philippines, Qatar, Rwanda, Singapore, Sri Lanka, Thailand, Tunisia, Uzbekistan and Venezuela.

Further information about CERN's international relations can be found at http://cern.ch/international-relations (site last visited 3/8/2013).

6. The Tevatron was a proton–antiproton collider. The antiproton is the "antiparticle" of the proton, and these have to be made at the laboratory, stored, "cooled," and re-injected into the machine to make collisions; see "Tevatron," http://en.wikipedia.org/wiki/Tevatron (site last visited 3/8/2013) and chapter 9.

7. See "LEP," http://en.wikipedia.org/wiki/LEP (site last visited 3/8/2013) and chapter 9.

8. The LHC experiments are discussed in further detail with many beautiful photographs at http://public.web.cern.ch/public/en/LHC/LHCExperiments-en.html (site last visited 3/7/2013). The ATLAS experiment is further discussed here: http://public.web.cern.ch/public/en/LHC/ATLAS-en.html (site last visited 3/7/2013). Quoting from the source:

> ATLAS is one of two general-purpose detectors at the LHC. It will investigate a wide range of physics, including the search for the Higgs boson, extra dimensions, and particles that could make up dark matter. ATLAS will record sets of measurements on the particles created in collisions—their paths, energies, and their identities. This is accomplished in ATLAS through six different detecting subsystems that identify particles and measure their momentum and energy. Another vital element of ATLAS is the huge magnet system that bends the paths of charged particles for momentum measurement. More than 2900 scientists from 172 institutes in 37 countries work on the ATLAS experiment (December 2009).

The CMS experiment is further discussed here: http://public.web.cern.ch/public/en/LHC/CMS-en.html (site last visited 3/7/2013). Quoting from this source:

The CMS experiment uses a general-purpose detector to investigate a wide range of physics, including the search for the Higgs boson, extra dimensions, and particles that could make up dark matter. Although it has the same scientific goals as the ATLAS experiment, it uses different technical solutions and design of its detector magnet system to achieve these. The CMS detector is built around a huge solenoid magnet. This takes the form of a cylindrical coil of superconducting cable that generates a magnetic field of 4 teslas, about 100,000 times that of the Earth. The magnetic field is confined by a steel "yoke" that forms the bulk of the detector's weight of 12,500 tonnes. An unusual feature of the CMS detector is that instead of being built in-situ underground, like the other giant detectors of the LHC experiments, it was constructed on the surface, before being lowered underground in 15 sections and reassembled. More than 2000 scientists collaborate in CMS, coming from 155 institutes in 37 countries (October 2006).

9. See "LHC Magnet Types," http://lhc-machine-outreach.web.cern.ch/lhc-machine-outreach/components/magnets/types_of_magnets.htm (site last visited 3/7/2013).

10. See "Magnet quench," http://en.wikipedia.org/wiki/Magnet_quench#Magnet_quench (site last visited 3/8/2013).

11. This is a heroic story and only attests to the profound success of the LHC project, yet it's hard to find a comprehensive historical record that captures the drama and intensity of this event. See "Large Hadron Collider," http://en.wikipedia.org/wiki/Large_Hadron_Collider (site last visited 3/7/2013). Quoting from this source:

> Problems occurred on 19 September 2008 during powering tests of the main dipole circuit, when an electrical fault in the bus between magnets caused a rupture and a leak of six tonnes of liquid helium. The operation was delayed for several months. It is currently believed that a faulty electrical connection between two magnets caused an arc, which compromised the liquid-helium containment. Once the cooling layer was broken, the helium flooded the surrounding vacuum layer with sufficient force to break 10-ton magnets from their mountings. The explosion also contaminated the proton tubes with soot. This accident was thoroughly discussed in a 22 February 2010 Superconductor Science and Technology article by CERN physicist Lucio Rossi.

See also BBC News: http://news.bbc.co.uk/2/hi/in_depth/7632408.stm; Daily Mail online: http://www.dailymail.co.uk/sciencetech/article-1256966/Large-Hadron-Collider-Cern-finally-reach-power-2013—years-schedule.html#axzz2JfKcqlSt (sites last visited 3/8/2013).

12. This is more properly known as the "Englert-Brout-Higgs-Guralnik-Hagen-

Kibble Mechanism," after its many contemporaneous co-inventors: http://en.wikipedia.org/wiki/Robert_Brout, http://en.wikipedia.org/wiki/Fran%C3%A7ois_Englert, http://en.wikipedia.org/wiki/Gerald_Guralnik; http://en.wikipedia.org/wiki/C._R._Hagen, http://en.wikipedia.org/wiki/Peter_Higgs, http://en.wikipedia.org/wiki/Tom_W._B._Kibble (sites last visited 3/7/2013).

13. See "London equations," the first theory of superconductivity in 1935, http://en.wikipedia.org/wiki/London_theory. The theory was proposed by brothers Fritz and Heinz London, though subsequent theoretical refinements containing the germ of the Higgs mechanism were due to Fritz London: http://en.wikipedia.org/wiki/Fritz_London. This is further generalized in the Ginzburg–Landau Theory of Superconductivity, in 1950: http://en.wikipedia.org/wiki/Ginzburg%E2%80%93Landau_theory. This is essentially a forerunner of the Higgs boson theory with the famous "Higgs potential" (sites last visited 3/8/2013).

14. See "Steven Weinberg," http://en.wikipedia.org/wiki/Steven_weinberg; "A Model of Leptons," *Phys. Rev. Lett.* 19, nos. 1264–66 (1967) was based upon Sheldon Glashow, "Partial Symmetries of the Weak Interactions," *Nucl. Phys.* 22, no. 579 (1961). This marks the beginning of the modern Standard Model era of particle physics. In Steven Weinberg, "The Crisis of Big Science," *New York Times Book Review* (2012), http://www.nybooks.com/articles/archives/2012/may/10/crisis-big-science/ (sites last visited 3/8/2013). Weinberg comments:

> In his recent book, *The Infinity Puzzle* (Basic Books, 2011), Frank Close points out that a mistake of mine was in part responsible for the term "Higgs boson." In my 1967 paper on the unification of weak and electromagnetic forces, I cited 1964 work by Peter Higgs and two other sets of theorists. This was because they had all explored the mathematics of symmetry-breaking in general theories with force-carrying particles, though they did not apply it to weak and electromagnetic forces. As known since 1961, a typical consequence of theories of symmetry-breaking is the appearance of new particles, as a sort of debris. A specific particle of this general class was predicted in my 1967 paper; this is the Higgs boson now being sought at the LHC.
>
> As to my responsibility for the name "Higgs boson," because of a mistake in reading the dates on these three earlier papers, I thought that the earliest was the one by Higgs, so in my 1967 paper I cited Higgs first, and have done so since then. Other physicists apparently have followed my lead. But as Close points out, the earliest paper of the three I cited was actually the one by Robert Brout and François Englert. In extenuation of my mistake, I should note that Higgs and Brout and Englert did their work independently and at about the same time, as also did the third group (Gerald Guralnik, C. R. Hagen, and Tom Kibble). But the name "Higgs boson" seems to have stuck.

15. See "Higgs boson," http://en.wikipedia.org/wiki/Higgs_boson (site last visited 3/8/2013).

16. Leon M. Lederman and Dick Teresi, *The God Particle: If the Universe Is the Answer, What Is the Question?* (Mariner Books, 2006); see also http://en.wikipedia.org/wiki/The_God_Particle:_If_the_Universe_Is_the_Answer,_What_Is_the_Question%3F (site last visited 4/8/2013).

CHAPTER 2. A BRIEF HISTORY OF THE BIG QUESTIONS

1. See "Democritus," http://en.wikipedia.org/wiki/Democritus; see also "Classical element" and references therein, http://en.wikipedia.org/wiki/Classical_element (sites last visited 3/8/2013).

2. Chemists write this as a chemical reaction: $CH_4 + 2O_2 \rightarrow CO_2 + 2H_2O$, which translates into "one methane (CH_4) and two oxygen molecules ($2O_2$) react to make one carbon dioxide (CO_2) and two water molecules ($2H_2O$)." Note that the total number of oxygen *atoms* on the left side of the reaction is $2 \times 2 = 4$, since there are two O_2 molecules, each containing two O atoms; the total afterward is $2 + 2 = 4$, since 2 O atoms enter the CO_2 molecule and two O atoms are present in the two water molecules.

3. See our book *Quantum Physics for Poets* (Amherst, NY: Prometheus Books, 2010).

4. See "Energy," http://en.wikipedia.org/wiki/Energy, "Electron volt," http://en.wikipedia.org/wiki/Electronvolt (sites last visited, 4/2/2013).

In discussing atoms and subatomic things, we use a very tiny quantity of energy, the *electron volt*, or *eV*. One electron volt is the energy that *a one-volt battery expends pushing a single electron* (the fundamental particle that orbits all atoms) *through an electric circuit*.

Most biological chemical reactions involve delicate, weaker chemical bonds, usually less than an electron volt per bond. Much of the entire stratum of biological processes is in the smaller energy range, around 0.1 eV. In contrast, the explosion of acetylene gas with oxygen liberates the energy of a triple carbon chemical bond—about 10 eV per acetylene molecule, and that is found to make a very loud bang—a shock wave in the surrounding air due to the release of chemical energy in the kinetic form. Indeed, high explosives such as TNT or fracking explosives can contain even higher energy bonds. Thus, around 10 eV, we enter the stratum of high-energy chemistry. By contrast, a proton and neutron can form a "deuteron," which is the simplest compound nucleus, the nucleus of deuterium, or "heavy hydrogen," and the energy released by this nuclear binding is about 2.23 MeV (million eV); this is a small nuclear binding energy. See http://en.wikipedia.org/wiki/Deuterium (site last visited 4/2/2013).

The conversion to larger energy units used in everyday engineering and macrophysics shows just how tiny this is. In the meter-kilogram-second system (mks), we use the energy unit "joule," which is one watt of power for one second. A 60-watt light is consuming *60 joules of energy per second.* This is the electrical power (energy per time) consumed by the lightbulb. An automobile driving down the road at 50 mph typically has a kinetic energy (energy of motion) of 450,000 joules.

The conversion of joules to eV is 1 joule = $6.24150974 \times 10^{18}$ *eV*. (That's approximately 6 followed by eighteen zeros.) The electron volt is a tiny unit of energy more appropriate for single atoms or particles, while joules are used for large macroscopic assemblages of atoms, such as mechanical engineering or electrical power applications.

Einstein's famous formula $E = mc^2$ implies that any particle at rest has an energy, E, that is equivalent to its inertial mass, m. We can therefore use energy units to quantify an elementary particle's mass. For example, the proton mass, which is $1.67262158 \times 10^{-27}$ kilograms, can be restated in terms of electron volts, using the conversion 1 GeV/c^2 = 1.783×10^{-27} kg, so a proton mass is 0.938 GeV/c^2 (1 GeV = 1 billion eV; 1 MeV = one million eV).

We'll often drop the "/c^2" and make a rough approximation, taking the proton and neutron masses to be approximately 1 GeV/c^2, abbreviated to 1 GeV. The electron mass, likewise, is 0.511 MeV/c^2, abbreviated to 0.5 MeV.

5. Chemists write this as a chemical reaction: $CH_4 + 2O_2 \rightarrow CO_2 + 2H_2O$ + heat, which translates into "one methane (CH_4) and two oxygen molecules ($2O_2$) react to make one carbon dioxide (CO_2) and two water molecules ($2H_2O$)." Note that the total number of oxygen *atoms* on the left side of the reaction is 2 × 2 = 4, since there are two O_2 molecules, each containing two O atoms; the total afterward is 2 + 2 = 4, since 2O atoms enter the CO_2 molecule and two O atoms are present in the two water molecules.

6. See "J. J. Thompson" and references therein, http://en.wikipedia.org/wiki/J._J._Thomson (site last visited 3/10/13).

7. See "Ernest Rutherford," http://en.wikipedia.org/wiki/Ernest_Rutherford (site last visited 3/10/13). "Good with his hands (unlike his mentor J.J. Thompson) and contemptuous of the head-in-the-clouds theoretical physicists, Rutherford was famous amongst his post-docs for acerbic quotes like the following: 'Oh, that stuff (Einstein's relativity). We never bother with that in our work,'" quoted in David Wilson, *Rutherford. Simple Genius* (Hodder & Stoughton, 1983); Richard Reeves, *A Force of Nature: The Frontier Genius of Ernest Rutherford* (New York: W. W. Norton, 2008).

8. Ibid.

9. Ibid.

10. Jan Faye, *Niels Bohr: His Heritage and Legacy* (Dordrecht: Kluwer Academic Publishers, 1991). See "Niels Bohr," http://en.wikipedia.org/wiki/Niels_Bohr (site last visited 3/10/13).

11. See "Cosmic ray," http://en.wikipedia.org/wiki/Cosmic_ray (site last visited 4/2/2013).

12. See "Nuclear physics," http://en.wikipedia.org/wiki/Nuclear_physics, "Atomic

nucleus," http://en.wikipedia.org/wiki/Atomic_nucleus, "Protons," http://en.wikipedia.org/wiki/Proton, "Neutrons," http://en.wikipedia.org/wiki/Neutron. These are generically called "nucleons," http://en.wikipedia.org/wiki/Nucleons. All atoms are characterized by the number of protons in their nucleus, or their "atomic number," http://en.wikipedia.org/wiki/Atomic_number. The number of neutrons can vary, from zero for hydrogen, to well over 150 for heavy unstable atoms, etc. For a fixed number of protons, the number of neutrons can vary, leading to "isotopes," http://en.wikipedia.org/wiki/Isotopes (all sites last visited 3/10/13).

13. See "Hideki Yukawa," http://en.wikipedia.org/wiki/Hideki_Yukawa, "Pion," http://en.wikipedia.org/wiki/Pion, "Strong force," http://en.wikipedia.org/wiki/Strong_force (all sites last visited 3/10/13).

The pion, like all strongly interacting particles, is actually a composite particle, made of a light quark (either "up" or "down") with a light anti-quark (either "anti-up" or "anti-down").

14. See "Muon," http://en.wikipedia.org/wiki/Muon (site last visited 3/10/13). Throughout this book we'll be placing emphasis on the muon because it is a truly point-like elementary particle (lepton) that has played a key role in unraveling the mystery of mass. It may also provide an avenue to the next generation of ultra-high-energy particle colliders, i.e., the Muon Collider. See http://en.wikipedia.org/wiki/Muon_collider (site last visited 3/10/13). Muons at rest live only about two millionths of a second (2×10^{-6} seconds). But because of a remarkable effect in relativity, as particles approach the speed of light, time slows down for them, and their lifetimes are extended; the muons are produced with energies that are hundreds of times their masses, and they can live hundreds of times longer, and this gets them down to the surface of the earth for detection.

15. See "Manhattan Project," http://en.wikipedia.org/wiki/Manhatten_project, and "Atomic bombings of Hiroshima and Nagasaki," http://en.wikipedia.org/wiki/Atomic_bombings_of_Hiroshima_and_Nagasaki (sites last visited 3/10/13).

CHAPTER 3. WHO ORDERED THAT?

1. See "Pion," http://en.wikipedia.org/wiki/Pion (site last visited 3/10/13). The positively charged π^+ is the "antiparticle" of the negatively charged π^-. Antimatter was theoretically predicted by the young genius Paul Dirac in 1926, who merged quantum theory together with Einstein's theory of special relativity, as we discuss this in chapter 9. A few years after Dirac's prediction, the antielectron, dubbed the "positron," was discovered in experiment. The theoretical prediction of antimatter is considered to be one of the greatest triumphs of twentieth-century physics. Each particle has a corresponding antiparticle species in nature. In some cases, like the photon and the neutral π^0, a particle can be its own antiparticle (we call such a particle "self-conjugate"; only electrically neutral particles can be self-conjugate).

2. The *Wikipedia* article on the muon reads as though we wrote it: http://en.wikipedia.org/wiki/Muon (site last visited 3/13/13).

3. The *Wikipedia* article on I. I. Rabi (pronounced "Robby"), a Nobel laureate, mentor of Leon Lederman, and a great and influential scientist, is well worth taking time to read: http://en.wikipedia.org/wiki/Isidor_Isaac_Rabi (site last visited 3/13/13). We quote the moving last paragraph from this article: "Rabi died at his home in Riverside Drive, Manhattan, from cancer, on 11 January 1988. In his last days, he was reminded of his greatest achievement in a poignant fashion when his physicians examined him using magnetic resonance imaging, a technology that had been developed from his ground-breaking research on magnetic resonance. 'I saw myself in that machine,' he remarked, 'I never thought my work would come to this.'"

4. See, e.g., the article on the Standard Model: http://en.wikipedia.org/wiki/Standard_model (site last visited 3/13/13). There are also many books and documentaries describing the history of particle physics and the Standard Model and beyond.

5. See "Muon-catalyzed fusion," http://en.wikipedia.org/wiki/Muon_catalyzed_fusion (site last visited 3/13/13).

6. See the Fermilab site for the Muon Collider: http://www.fnal.gov/pub/muon_collider/. Searching online for "Muon Collider" pulls up stuff like: http://www.dvice.com/archives/2012/07/muon-collider-c.php (sites last visited 3/13/13).

7. See the "History" section of the "Muon" entry: http://en.wikipedia.org/wiki/Muon#History (site last visited 3/13/13).

8. So, in summary, the muons are *not the primary* cosmic rays, and they aren't even the secondary ones (the pions are secondaries) but rather they are the *tertiary products* of decaying pions produced in the high-up-in-the-atmosphere cosmic ray collisions! The muons are deeply penetrating because they don't have strong interactions and make it down to the surface of the earth where they could be detected, but the pions are reabsorbed in collisions with nuclei, and they have a shorter lifetime, so they don't make it to lower levels of the atmosphere.

9. What we usually quote as a particle "lifetime" is actually the "*e-folding time*," which is slightly longer than the half-life. That is, the number of particles, at time t, is proportional to $e^{-t/\tau}$ where τ is the e-folding time; this is equivalently to $2^{-t/\tau'}$ where τ' is the *half-life*, and $\tau = \tau'/\ln(2) = 1.44\ \tau'$.

10. See the "History" section of the "Pion" entry: http://en.wikipedia.org/wiki/Pion#History. In 1948 the first "artificially produced" pions were created by the cyclotron at the University of California, Berkeley. The charged pions that are directly produced in the upper atmosphere decayed into the muons that were ultimately detected in 1937 at the surface of the earth. Since muons do not interact strongly with protons of neutrons they are able to penetrate matter very deeply and arrive at the surface of the earth (with some help from Einstein's time dilation). In fact, you must go rather deep into the earth to shield

yourself from having any muons hit you, as they can penetrate to great depths through rock. The fact that all of this deep penetration through matter happens tells us that muons are not strongly interacting particles—i.e., they were not the pions predicted by Yukawa.

Muons do have electric charges, and when a particle that has an electric charge passes through matter, it gradually loses energy by virtue of the electromagnetic interaction. Typically, it slightly scatters off of an electron in an atom and is slightly accelerated, causing it to radiate a photon.

Photons are radiated from charged particles when they accelerate; that's how light is produced by a chemical reaction, such as burning. The reaction rapidly accelerates electrons as they collide with one another. This is also the principle of radio transmission. An antenna is a wire in which electrons are accelerated, causing them to emit the low-energy photons that are radio waves. This photon radiation has a fancy Germanic name: *bremsstrahlung*. An example of this is a dental X-ray machine, where accelerated electrons, like those in an old TV picture tube, collide with a metal target. In the target the electrons instantly decelerate, losing all of their energy by collisions with other electrons and atomic nuclei. Whenever the electron collides, it radiates a bremsstrahlung photon. If the initial electron energy is high, the resulting emitted photons will also be energetic, and this can make X-rays, or gamma rays, etc. We'll henceforth call this "brem radiation" or "brem photons" like our colleagues at the lunch table at Fermilab do. You can use this particle physics slang to impress your significant other.

The amount of brem radiation depends upon how much the charged particle is instantaneously accelerated in a collision. An electron is a very light particle, having a tiny mass of only 0.511 MeV. It can therefore get bumped around and jiggle and bounce, thus accelerating a lot in collisions. Think of box of Ping-Pong® balls rattling around in a truck on a bumpy road. The Ping-Pong ball is such a featherweight that it can't sit still in a bumpy environment. Electrons therefore produce a lot of brem photons when they pass through matter because they bounce around and recoil in collisions so much. Electrons therefore quickly lose all of their energy when they enter matter. If the muon were as light as an electron, it would never reach the surface of the earth after being produced by a primary collision ten miles up in the sky. A muon, on the other hand, is 200 times heavier than an electron. Muons are like bowling balls compared to electrons. They don't accelerate or recoil much as they pass through matter and collide with atoms and electrons. So, muons tend not to produce so many brem photons. The muons therefore don't lose energy very much when they pass through matter. This is one of the reasons muons can make it to the surface of the earth and go considerable distance down through rock once they're produced some ten miles up in the atmosphere from decaying pions.

11. The pions are composite particles, made of quarks, as are the proton and neutron. All quark containing particles have strong interactions with one another (see the Appendix). Muons, however, are leptons, and leptons do not experience this strong force. This may be,

in a larger sense, the focus of the question "Who ordered that?"—it remains a mystery why we have these two particular classes of particles: quarks with their strong interactions that bind them eternally into protons and neutrons and other strongly interacting particles, and leptons that just don't experience these strong forces.

A neutral pion contains a quark and an antiquark, and this is just a particle annihilating (quark) with an antiparticle (antiquark) inside the pion. All of the everyday nuclear matter in our world is composed of combinations of the two quarks, up and down. These objects are distinguished by their electric charges and their masses. We always define the electron to have an electric charge of -1. In these units, the up quark (u) has an electric charge of +2/3, and the down quark (d) an electric charge of -1/3. The proton is therefore not an elementary particle but rather a composite particle built of three quarks in the pattern $u + u + d$. Adding up the electric charges of the constituent quarks, we see that the proton charge is +1. Similarly, the neutron is composed of $u + d + d$, and the corresponding electric charge combination is 0. See the Appendix for more details.

12. Lewis Carroll, *Alice's Adventures in Wonderland* and *Through the Looking-Glass and What Alice Found There*, illustrations by John Tenniel (London: Macmillan and Co., 1866 & 72, 1866), 2 volumes. The marvelous illustrations are cataloged here: https://sites.google.com/site/lewiscarrollillustratedalice/ (site last visited 3/7/13).

13. Ibid.

14. This account appears in Leon M. Lederman and Dick Teresi, *The God Particle: If the Universe Is the Answer, What Is the Question?* (Mariner Books, 2006) and in Leon M. Lederman and Christopher T. Hill, *Symmetry and the Beautiful Universe* by (Amherst, NY: Prometheus Books, 2008).

15. Note that Lederman et al. were using positive muons that are always right-handed (R) in positively charged pion decay; the positively charged muon is actually the antiparticle of the muon, which, like the electron, is negatively charged. If Leon did the experiment with negatively charged muons they would all be left-handed, or L, in the negative pion decay. The chirality of a particle is always the opposite of the chirality of the antiparticle (as discussed in chapter 9, the anti-L particle is the *absence of* a negative energy L particle, or a "hole" in the vacuum, so the hole would be R).

16. See "Parity" (physics), http://en.wikipedia.org/wiki/Parity_%28physics%29, "T. D. Lee," http://en.wikipedia.org/wiki/T_D_Lee, "C. N. Yang," http://en.wikipedia.org/wiki/C_N_Yang (sites last visited 3/13/13).

17. See "Chien-Shiung Wu," http://en.wikipedia.org/wiki/Madame_Wu (site last visited 3/13/13).

CHAPTER 4. ALL ABOUT MASS

1. Neil DeGrasse Tyson quote: https://www.facebook.com/pages/Higgs-Boson/384970858230425.

2. "'Yeah, the Higgs boson is getting a lot of attention, but there are a lot of lower-profile bosons that are worth checking out if you get the chance'—Tamara Farrar—Unemployed," http://www.theonion.com/articles/2012-in-technology,30782/ (site last visited 3/8/2013).

3. Perhaps to draw some sympathy for the plight of providing a more precise and universal definition in the text, we provide here a list and discussion of other forms of energy from our book *Symmetry and the Beautiful Universe* (Amherst, NY: Prometheus Books, 2007), chap. 2:

> *Potential energy* is energy that is stored in an object or system, which is ready upon its release to make other objects move. For example, a compressed spring has potential energy that can launch a toy dart from a child's rubber tipped dart gun, or help to hoist open a garage door, or to run an old fashioned wind-up watch for days. This energy in a spring is actually the energy of deformation of the lattice of iron alloy (steel) atoms within the material as they are twisted slightly away from their normal relaxed pose. There can be many forms of potential energy. For example, a bank of snow sitting atop a mountain has gravitational potential energy, ready at an instant to fall and convert to kinetic energy of motion. Gasoline, or other fuels, contains chemical potential energy waiting to be released by the chemical reaction of oxidation (burning).
>
> *Chemical energy* is created (or consumed) as various substances can undergo a vast array of chemical reactions that produce or consume energy. The precise form that the chemical energy takes depends upon the reaction. One rather common example is the burning of coal, oil, wood, or other substances which have a high percentage of carbon. "Burning" is the act of combining carbon with oxygen (a gas that is conveniently and generously supplied by the atmosphere). The basic reaction is $C + O_2 \rightarrow CO_2 + Q$. Here Q is a symbol for energy, which includes particles of light (known as photons, or equivalently, the particles that make up electromagnetic radiation), and the rapid motion (kinetic energy) of the resulting molecules after burning. In other words, carbon combines with oxygen to produce carbon-dioxide plus energy.
>
> *Thermal energy* is the high speed random motion of atoms, molecules, or other particles in a gas or other material. In a wood fireplace, the rapidly moving molecules, products of the combustion, collide with other molecules, such as the surrounding air, giving them kinetic energy, which propagates the heat outward into the room in a process called convection. The photons stream out into the room, as well as thermal radiation, producing radiant heat. The pleasurable sensa-

tion of a warm fire in a fireplace is nothing more than a bath of faster moving air molecules and photons.

Electrical energy is yet another form. In its simplest form, this is just the kinetic energy of the flow of electrons (an electric current) through a wire, through certain liquids, or in free space as in a vacuum tube (such as a cathode ray tube, a TV picture tube) or in a particle accelerator of electrons. If the wire has a large electrical resistance, then the electrons collide with the atoms in the wire, losing their energy, causing the wire atoms to move. The wire thus becomes hot, as in a toaster or an electric oven broiler. This is called electrical resistance, and leads to the loss of electrical energy.

The tricky thing about the book-keeping of electrical energy, however, is that an electron can radiate, or emit, a photon, the particle of light, through the process: The electron, after the emission of the photon has less energy than the electron before the "emitted photon," has been created, carrying off a certain amount of energy. This can also go in reverse where the initial electron absorbs the photon and the final electron has gained the photon's energy. This is the fundamental process in nature that defines electromagnetism. Photons can be stored in an "electromagnetic field," as a kind of photon soup, which itself contains the energy of the photons. So, energy in the basic processes of electricity and magnetism is difficult to keep track of, involving continual swapping back and forth between electrons and photons. Chemical energy, when examined microscopically, is really electrical energy within atoms and molecules. As we have indicated, energy is a precisely defined concept in physics. It is a useful concept because it is conserved in all processes. If we have a large box inside of which all possible things can happen, such as springs that can compress and expand, various bodies can fall and bounce, water can flow, chemicals react, objects burn, atomic nuclei disintegrate, etc., and through all of this there is one number that stays the same—the total energy.

We living organisms are also engines. Our bodies are consuming energy to sustain our metabolism, ergo our lives. Here we measure energy in "food calories," usually designated with the upper case "C" as in Calorie. A typical (lean) person in America eats about 2000 Calories per day. To convert this into joules we multiply by (approximately) 4,200, hence the average lean person is consuming about 8,400,000, or 8.4 million joules of food energy per day! In a day there are 24 hours and 60 minutes per hour, and 60 seconds per minute, that is, 86,400 seconds total in a day. Therefore, the average person consumes energy, and burns off the equivalent energy, at an average rate of about 8,400,000/86,400 = 97 watts. Therefore, each of us, as living, functioning, metabolizing beings, is approximately equivalent to a 100 watt light bulb in our metabolic power consumption. For comparison, the sun produces at ground level, a power of about 100 watts per square meter, averaged over a sunny day. So, a 30 meter by 30 meter solar collector, about the size of a large roof, with an efficiency of 10%, would be required for every person in our society to obtain all the presently required power from the sun.

Power: The "time rate" at which energy is produced or consumed or converted is called *power*. One can think of power as a kind of speed, if one thinks of energy as a kind of distance. If you want to take a trip somewhere, you must travel a certain distance. How fast you do this depends upon your speed. The greater the speed, the shorter the time for the trip. Likewise, you may want to consume a certain amount of energy to perform a task, such as mowing the lawn. How fast you perform the task determines the power you require, the time rate at which you consume the energy. The more power you consume, the shorter the time the task will take. Note that the power is not a fixed or conserved quantity because we can speed up or slow down the rate at which we perform the task. The total energy, on the other hand, is fixed, like the total distance traveled for a given trip.

We might ask, "If a moving car consumes energy, where did the all that energy go?" If you ask that question, then you have indeed learned our all-important lesson about energy—energy is conserved and cannot be created or destroyed—it therefore must have gone somewhere else. In the case of our car, the kinetic energy is lost through friction between mechanical parts, into heating the engine, through the sound energy the car produces, into the energy content of the air that the car moves around as it travels down the highway, and into the energy of heating and compressing and deforming the tires as they spin around. However, most of the wasted energy goes into heat, increasing the speed of the molecules of water (engine coolant), tires, the road, etc. Since this is chaotic and random molecular motion, it is virtually impossible to usefully recover this energy.

4. In relativity we find the energy and momentum of the moving particle

Einstein: $E = \dfrac{mc^2}{\sqrt{1-v^2/c^2}} \quad \vec{p} = \dfrac{m\vec{v}}{\sqrt{1-v^2/c^2}}$

Newton: $E = \dfrac{1}{2}mv^2 \quad \vec{p} = m\vec{v}$

where we have written the Newtonian expressions for comparison. We can see the "Newtonian limit" and the nonzero rest energy implied for zero velocity, $v = 0$, where the Einstein formulas go into

$$E = mc^2 + \frac{1}{2}mv^2, \text{and } \vec{p} = m\vec{v}.$$

In special relativity, we can never get the speed of a massive particle (one with nonzero inertial mass m) to equal the speed of light. As $|\vec{v}| \rightarrow c$ the momentum and the energy become infinite. It therefore would require an infinite energy to accelerate a proton to the speed of light. $E = mc^2$ holds strictly for a particle at rest.

At the CERN LHC at full design energy we accelerate protons to seven trillion electron volts. The rest mass energy of a proton is about one billion electron volts. Hence, the LHC "boosts" a proton to have a "Lorentz factor" $\gamma = 1/\sqrt{1-v^2/c^2}$ of about 7,000. This means that approximately $(v/c)^2 = 0.99999998$, or that the LHC accelerates protons to 99.999999 percent of the speed of light.

How, then, can anything travel at the speed of light? We see that if we take $|\vec{v}| = c$ and also allow our particle to be massless, then the energy is actually indeterminate, that is, we get, $E = 0/0$. However, this allows for the possibility that a massless particle, something with no inertial mass, can have finite energy and momentum. If we look at the relationship between energy and momentum, we see that a massless particle must satisfy: $E = |\vec{p}|c$. Indeed, this describes the particles of light, the photons. Photons have absolutely no inertial mass, yet they carry energy and momentum as they travel through space. Photons travel forever at the speed of light. They cannot be at rest, or have a finite velocity less than c, for then their energy would be zero.

5. For a short yet poignant biography of Emmy Noether, see chapter 3 of *Symmetry and the Beautiful Universe* by Lederman and Hill. Also, the "conservation of energy" and its relationship to the "constancy of the laws of physics in time" are explained by the amusing example of the "ACME Power Company" in chapter 2.

CHAPTER 5. MASS UNDER THE MICROSCOPE

1. To measure the mass of a water molecule, we start by looking up the mass of 1 mole of hydrogen, which is listed on the Periodic Table of the Elements as about 1 gram. A mole of oxygen is listed as 16 grams. Therefore, an H_2O molecule has 18 grams per mole. A "mole" is a definite number of particles, about 6×10^{23} particles. This is "Avogadro's number," and determining this number makes the basic connection between aggregate matter and the number of atoms per unit mass comprising it (we're rounding things to one significant digit). So the mass of a single H_2O molecule is 18 grams/ (6×10^{23}), or 3×10^{-23} grams, hence 3×10^{-26} kilograms.

In this way, with more care, we can measure the mass of the proton, the neutron, and—with more effort—even the lowly electron, as well as masses of all sorts of atomic and subatomic things. We often quote these masses as an energy, using Einstein's $E = mc^2$. We did this earlier for the electron (about 0.5 million electron volts, or MeV, of mass in equivalent energy units) and our celebrant, the muon (about 105 MeV).

Sometimes you'll see a mass quoted in MeV/c^2, or GeV/c^2, which rewrites Einstein's formula as $m = E/c^2$ and installs the factor of $/c^2$. Most physicists simply use units in which c = 1 and Planck's constant h-bar = $h/2\pi = 1$, and then use electron volts, eV, as the measure of mass, distance, and time. For example, 1 GeV (Giga electron volts) corresponds to $(0.2 \times 10^{-13}\ \text{cm})^{-1}$.

2. See "Galaxy," http://en.wikipedia.org/wiki/Galaxy (site last visited 4/1/2013).

3. See chapter 1, note 13; see also "Superconductivity," http://en.wikipedia.org/wiki/Superconductivity,"Ginzburg–LandauTheory,"http://en.wikipedia.org/wiki/Ginzburg%E2%80%93Landau_theory, "John Bardeen," http://en.wikipedia.org/wiki/John_Bardeen, "Leon Cooper," http://en.wikipedia.org/wiki/Leon_Cooper, "Robert Schrieffer," http://en.wikipedia.org/wiki/Robert_Schrieffer, "V. Ginzburg," http://en.wikipedia.org/wiki/Vitaly_Lazarevich_Ginzburg, "Lev Landau" http://en.wikipedia.org/wiki/Lev_Landau (sites last visited 4/1/2013).

In a superconductor the vibrations of the lattice of atoms that make up the material interact with the electrons. This interaction is very delicate, but at ultra-cold temperatures, it causes the electrons to pair up, forming little particles that are like "two-electron atoms" and that behave quite differently than unpaired electrons. These are called Cooper pairs, and they form a kind of quantum soup inside of the superconductor. This quantum soup is the structure of the vacuum itself, the state of lowest energy, inside of a superconductor. When a photon then enters the superconductor it interacts with the soup. This causes the photon to blend with the soup and behave exactly like a photon that has a mass. The massive photon can be brought to rest in the superconductor, but in fact it is really the original massless photon and soup combination that is brought to rest.

4. See "Jeffrey Goldstone," http://en.wikipedia.org/wiki/Jeffrey_Goldstone, "Giovanni Jona-Lasinio," http://en.wikipedia.org/wiki/Giovanni_Jona-Lasinio, "Yoichiro Nambu," http://en.wikipedia.org/wiki/Yoichiro_Nambu (sites last visited 4/2/2013).

5. This is a quantum phenomenon. The rapid flip-flop between L and R chiralities is associated with a rapid oscillating behavior of the muon at rest. This was first realized when people considered the resting state solutions to Dirac's equation that described quantum states of electrons, muons, etc. in a manner consistent with relativity. It was given a fancy German name: "Zitterbewegung."

One might be worried that the L-R-L-R oscillation involves a rapid change in momentum from east to west, back to east, etc. For a localized particle, one at rest, this rapid fluctuation in momentum can be viewed as consistent with Heisenberg's uncertainty principle, $\Delta p\, \Delta x < h/\, 2\pi$. The "point-like" resting massive particle can be viewed as localized to within $\Delta x = h/2\pi mc$ (the "Compton wavelength") and can therefore fluctuate in momentum by $\Delta p = mc$.

CHAPTER 6. THE WEAK INTERACTIONS AND THE HIGGS BOSON

1. Leon Lederman, Melvin Schwartz, and Jack Steinberger received the Nobel Prize in Physics in 1988 for their research that revealed the "flavors" of neutrinos. The neutrinos are paired with corresponding charged leptons. There are therefore three charged leptons: electron, muon, and tau, and three associated neutrinos:

$$e^- \leftrightarrow \nu_e; \qquad \mu^- \leftrightarrow \nu_\mu; \qquad \tau^- \leftrightarrow \nu_\tau$$

2. Heisenberg's uncertainty principle: The uncertainty principle implies that if we try to localize any particle in space within a very small region of distance, Δx, the uncertainty in position, then the uncertainty in the x-component of the momentum of the particle, Δp_x, will grow larger, becoming at least as big as $\Delta p_x \geq h/2\pi\Delta x$. Similarly, if we want to localize some event in a system within a tiny time interval, Δt, then we will necessarily disturb the system and cause a range in its energy of ΔE, where $\Delta E \Delta t \geq h/2\pi$, so the smaller we make Δt, the larger becomes ΔE, as $\Delta E \geq \hbar/\Delta t$. The atomic orbitals of electrons have a typical size in most atoms of roughly $\Delta x \approx 10^{-10}$ meters in any given direction in space. Therefore, electrons must, by the uncertainty principle, have a range of momentum within their orbitals that is as large as $\Delta p_x \geq \hbar/\Delta x$, hence, $\Delta p_x \approx 10^{-24}$ kilogram-meter/second. Electrons move in their orbitals with velocities that are much less than c (i.e., they are *non-relativistic*), and the electron mass is known to be $m_e \approx 9.1 \times 10^{-31}$ kilograms. Therefore, we can estimate the typical electron kinetic energies to be of order, $E \approx (\Delta p_x)/2m_e \approx 6 \times 10^{-19}$ joules, or about 3.8 electron volts (1 electron volt = 1.6×10^{-19} joules; we have done a lot of "rounding off" to do this "back-of-the-envelope" estimate). The force that holds the electrons in their orbitals must therefore provide a negative potential energy that exceeds, in magnitude, this result. This is the electromagnetic force, and the typical scale of the *binding energies* of electrons in an atom (the energy we must supply to liberate them) is of this order, ranging over about 0.1 to 10 electron volts. In fact, this is the typical energy scale of all chemical processes, and it contains the typical energies of visible light photons.

A "quantum fluctuation" is a bit like a "thermal fluctuation." It is physically possible for a thermal system, like a hot gas in a room, to suddenly fluctuate in density and pressure—even the extreme fluctuation of all of the gas momentarily condensing onto the floor then evaporating back into the room is physically possible, but such a thing is ultra-ultra rare. Note that a top quark is heavy enough that the top can directly decay, converting to a b-quark and a W^+, without requiring the uncertainty principle.

3. In any physical process, when a direction in space (called a vector) becomes correlated with a spin or a magnetic field (which is a "pseudovector"; the mirror image of a pseudovector is opposite that of a vector), then there is parity violation, since in the looking-

glass house the correlation would be opposite; i.e., for Leon's experiment, in the looking-glass house the electrons would come out in the opposite direction of the muon spin. In Madame Wu's version of the experiment, the electrons coming out of ^{60}Co decay were aligned with the direction of the magnetic field used to align the spins of the nuclei.

4. The relationship is simple math, if you have had calculus: the electric and magnetic fields are the particular derivatives of the gauge field in space and time.

5. Sheldon Glashow, "Partial Symmetries of the Weak Interactions," *Nuclear Physics* 22 (1961): 579–88.

6. Steven Weinberg, "A Model of Leptons," *Physical Review Letters* 19 (1967): 1264–66.

7. Gerard 't Hooft and Martinus Veltman, "Regularization and Renormalization of Gauge Fields," *Nuclear Physics* B44 (1972): 189–213.

8. In particular, fermions are spin-1/2, or "half-integer" spins. See Leon M. Lederman and Christopher T. Hill, *Quantum Physics for Poets* (Amherst, NY: Prometheus Books, 2011) or the discussion of spin in the Appendix.

9. See "Satyendra Nath Bose," http://en.wikipedia.org/wiki/Satyendra_Nath_Bose (site last visited 1/23/2013).

10. This is a complicated phenomenon that involves the interactions of the electrons in the atoms—for most magnetic materials the atoms align but cancel the overall field to zero. Iron prefers a ground state in which there is an exact common alignment, and we get a big magnetic field emanating from the iron.

CHAPTER 7. MICROSCOPES TO PARTICLE ACCELERATORS

1. This news bulletin was from CNN Tech: http://articles.cnn.com/2009-11-21/tech/cern.hadron.collider_1_large-hadron-collider-lhc-cern?_s=PM:TECH (site last visited 1/23/2013).

2. Some early history of microscopes can be found here: http://en.wikipedia.org/wiki/History_of_optics; http://en.wikipedia.org/wiki/Magnifying_glass ; http://www.history-of-the-microscope.org/ (sites last visited 6/21/2013).

3. See "Zacharias Jannsen" and references therein, http://en.wikipedia.org/wiki/Sacharias_Jansen; "History of the Microscope" http://www.history-of-the-microscope.org/hans-and-zacharias-jansen-microscope-history.php (sites last visited 1/23/2013).

4. Miscellaneous references on the history of microscopes: http://inventors.about.com/od/mstartinventions/a/microscope.htm (site last visited 1/23/2013); R. M. Allen, *The Microscope* (New York: D. Van Nostrand Company, Inc., 1940); *The* S. Bradbury, *Evolution of the Microscope* (Oxford: Pergamon Press, 1967); W. G. Hartly, *The Light Microscope* (Oxford: Senecio Publishing Company, 1993). See also "Telescope," http://en.wikipedia.org/wiki/

Telescope; Henry C. King and Harold Spencer Jones, *The History of the Telescope* (Courier Dover Publications, 2003).

5. See "Antonie van Leeuwenhoek," http://en.wikipedia.org/wiki/Van_Leeuwenhoek (site last visited 1/23/2013); Alma Smith Payne, *The Cleere Observer: A Biography of Antoni van Leeuwenhoek* (London: Macmillan, 1970).

6. Anton van Leeuwenhoek, Letter of June 12, 1716. The letter and short biography can be found here: http://www.ucmp.berkeley.edu/history/leeuwenhoek.html (site last visited 1/23/2013).

7. See "Robert Hooke" and references therein, http://en.wikipedia.org/wiki/Robert _Hooke (last visited 1/23/2013).

8. A beam of light bends as it obliquely hits water or glass. This bending of light by transparent materials is called refraction. The amount of refraction is controlled by the "index of refraction" of the medium the light is exiting (e.g., air) and that the light is entering. The index of refraction varies with light wavelength. This is the basis of the phenomenon of a glass prism that splits the white light into its spectral constituents: Red-Orange-Yellow-Green-Blue-Indigo-Violet (ROY G. BIV). White light is therefore composed of equal amounts of the different colors of light. We can take the colors of light and combine them to make white light. The chromatic aberration is mainly the prism effect of the glass lens.

9. See "Joseph Jackson Lister," http://en.wikipedia.org/wiki/Joseph_Jackson_Lister; see also "Lens," under "Optics" heading," http://en.wikipedia.org/wiki/Lens (sites last visited 1/23/2013). Lister published his work in 1830 in a paper titled "On Some Properties in Achromatic Object-Glasses Applicable to the Improvement of the Microscope," submitted to the Royal Society.

10. Lenses and aberrations are thoroughly discussed here: http://en.wikipedia.org/wiki/Lens_%28optics%29#Compound_lenses, and here: http://en.wikipedia.org/wiki/Compound_lens (sites last visited 4/10/2013).

11. D. Edwards and M. Syphers explain this and much of optics with some simple matrix algebra in an elegant book: *An Introduction to the Physics of High Energy Accelerators (Wiley Series in Beam Physics and Accelerator Technology)* (Wiley, 1992), pp. 60–65.

12. Let's review the physics of a wave: Consider a long traveling wave as it moves through space. We can visualize this as a freight train moving by as we are stopped at a railroad crossing. A traveling wave is sometimes called a wave train, with many sequential crests and troughs of the train as it traverses space. Such a wave is described by three quantities: its frequency, its wavelength, and its amplitude. The wavelength is the distance between two neighboring troughs or crests of the wave. The frequency is the number of times per second the wave undulates up and down through complete cycles at any fixed point in space. If we think of the wave as a long freight train, its wavelength is then the length of a boxcar. Its frequency is the number of box cars per second passing in front of us as we patiently wait for the train to pass. The speed of the traveling wave is therefore the

length of a boxcar divided by the time it takes to pass, or (speed of wave) = (wavelength) times (frequency). Thus, knowing the speed, the wavelength and frequency are inversely related, or (wavelength) = (speed of wave) divided by (frequency) and (frequency) = (speed of wave) divided by (wavelength). The amplitude of the wave is the height of the crests, or the depth of the troughs, measured from the average. That is, the distance from the top of a crest to the bottom of a trough is twice the amplitude of the wave, and it can be thought of as the height of the boxcars. For an electromagnetic wave, the amplitude is the strength of the electric field in the wave. For a water wave, twice the amplitude is the distance that a boat is lifted from the trough to the crest as the wave passes by. Figure 2.1 says it all. The color of a visible light wave was understood in the nineteenth-century Maxwellian theory of electromagnetism to be determined by the wavelength (and, inversely, the by frequency). If we take the frequency to be small, we correspondingly find that the wavelength becomes large. Longer-wavelength visible light is red, while shorter-wavelength visible light is blue. For graphic display in color of the various wavelengths of light, see http://science-edu.larc.nasa.gov/EDDOCS/Wavelengths_for_Colors.html#blue (site last visited 3/26/2013).

13. See "Louis de Broglie," http://en.wikipedia.org/wiki/Louis_de_Broglie; see also "Davisson–Germer experiment," http://en.wikipedia.org/wiki/Davisson%E2%80%93 Germer_experiment (site last visited 4/10/2013).

14. According to Newton, the *magnitude* of the force of gravity exerted upon object **a** by object **b** is called F_{ab} and is given by the formula:

$$F_{ab} = \frac{G_N m_a m_b}{R^2}$$

where R is the separation between them. This is an example of an *inverse square law force*, that is, a force that falls off in magnitude, or strength, with distance, like $1/R^2$. The electric force between two stationary electric charges is also an inverse square law force.

In this formula m_a is the mass of object **a**, and m_b is the mass of object **b**. This means that the force of gravity is stronger between two very massive objects than it is between two very low-mass objects. For example, if **a** is the earth, we substitute $m_a = m_{Earth}$, and if **b** is the sun, we substitute $m_b = m_{Sun}$ into the formula. Thus, if we could somehow double the mass of the sun, holding everything else fixed, then the force of gravity that the earth would experience from the sun would become doubled, and the earth's orbit would change, becoming a tighter ellipse with a smaller average distance from the sun. Technically, the force is a *vector* and must therefore also have a direction. We could write a better formula that illustrates that, but words suffice. Object **a** experiences the force of gravity, with the magnitude we have written, but the force points as a vector at the direction of object **b**. And, by symmetry, object **b** experiences the same magnitude of force, which points in exactly the opposite direction, back to object **a**.

The quantity G_N in the numerator of the formula is a *fundamental constant*. Newton had to introduce this factor in order to specify the *strength* of the gravitational force. We call this Newton's gravitational constant or just Newton's constant, for short. G_N is measured from experiment and takes the value $G_N = 6.673 \times 10^{-11}$ (meters^3) / (kilograms seconds^2). We have quoted G_N in the meter-kilogram-second system of units. Indeed, we can write, in nonscientific notation, $G_N = 0.00000000006673$ (meters^3) / (kilograms seconds^2), and we see that G_N is a seemingly very small number. Gravity, despite its ubiquitous character in nature, is actually a very feeble force. To get a sense of this, we can estimate that the force of gravitational attraction between two fully loaded oil tankers that are ten miles apart is about the same as the force you feel holding a gallon of milk due to the pull of gravity by the entire earth. For more discussion of gravity, see our book *Symmetry and the Beautiful Universe* (Amherst, NY: Prometheus Books, 2007).

15. See "Charles-Augustin de Coulomb," http://en.wikipedia.org/wiki/Charles-Augustin_de_Coulomb, and "Coulomb's law," http://en.wikipedia.org/wiki/Coulomb%27s_law (sites last visited 3/26/2013).

16. See "Electric charge," http://en.wikipedia.org/wiki/Electric_charge. The electric field, which we call E, produces a force on the charge, which we call F, and the relationship between these is very simple, F = eE, or "force equals charge times electric field." A force causes a particle to accelerate. This was precisely expressed by Newton in his famous equation F = ma, or "force equals mass times acceleration." So, if we combine these two equations, we see that ma = eE, or a = eE/m, "the acceleration of the particle is proportional to charge times electric field divided by mass." See "Electric field," http://en.wikipedia.org/wiki/Electric_field; see also http://hyperphysics.phy-astr.gsu.edu/hbase/electric/elefie.html; http://www4.uwsp.edu/physastr/kmenning/Phys250/Lect03.html (sites last visited 1/23/2013). Quarks have fractional charges, up = +2/3 times e, and down = −1/3 times e, but quarks are always bound into strongly interacting particles, such as protons, neutrons, and π's such that the observed charges are integers (see Appendix).

17. The direction of an electric field and a *conventional* electric current always emanates from positive and points toward negative. This convention was adopted before the discovery that the electric charge of the electron is negative. So the actual flow of electrons is opposite to that of the electric field and the conventional current. See "Electric current," http://en.wikipedia.org/wiki/Electric_current (site last visited 5/4/2013).

18. Search online for "electron microscopes" and follow links to "images for electron microscopes." See "Electron microscope" and references therein, http://en.wikipedia.org/wiki/Electron_microscopes (site last visited 1/23/2013). Quoting from this source:

> According to Dennis Gabor, the physicist Leó Szilárd tried in 1928 to convince him to build an electron microscope, for which he had filed a patent. The German physicist Ernst Ruska and the electrical engineer Max Knoll constructed the prototype electron microscope in 1931, capable of four-hundred-power magni-

fication; the apparatus was a practical application of the principles of electron microscopy. Two years later, in 1933, Ruska built an electron microscope that exceeded the resolution attainable with an optical (lens) microscope. Moreover, Reinhold Rudenberg, the scientific director of Siemens-Schuckertwerke, obtained the patent for the electron microscope in May 1931. Family illness compelled the electrical engineer to devise an electrostatic microscope, because he wanted to make visible the poliomyelitis virus. The first practical electron microscope was constructed in 1938, at the University of Toronto, by Eli Franklin Burton and students Cecil Hall, James Hillier, and Albert Prebus; and Siemens produced the first commercial transmission electron microscope (TEM) in 1939. Although contemporary electron microscopes are capable of two million-power magnification, as scientific instruments, they remain based upon Ruska's prototype.

CHAPTER 8. THE WORLD'S MOST POWERFUL PARTICLE ACCELERATORS

1. See "Michael Faraday," http://en.wikipedia.org/wiki/Michael_Faraday (site last visited 3/26/2013). In the "opinion" of Snopes.com (the fact-checking website that has punched so many holes in the many idiotic opines of elected officials and various rancidly political distribution e-mails), this famous quote of Faraday's is undocumented hearsay: http://www.snopes.com/quotes/faraday.asp. However, the recipient of the comment, Gladstone, was supposedly Chancellor of the Exchequer, and not prime minister, according to Wikiquote: http://en.wikiquote.org/wiki/Michael_Faraday (sites last visited 3/26/13): "Faraday's reply to William Gladstone, then British Chancellor of the Exchequer (minister of finance), when asked of the practical value of electricity (1850), as quoted in *The Harvest of a Quiet Eye: A Selection of Scientific Quotations* (1977), p. 56." Snopes claims to discredit the quote because Gladstone was allegedly prime minister at the time of the remark, but in fact he did not hold that office until after Faraday's death. The fact that Gladstone was Chancellor of the Exchequer seems to undercut that part of the Snopes argument. Electricity had not developed to the cell phone–video camera stage in Faraday's era, so we'll never know who's right or who's wrong, but we do love the "quote."

2. Much more technical detail than we have space for can be found by perusing the Wikipedia entry for "Linear particle accelerator," http://en.wikipedia.org/wiki/Linear_Accelerator (site last visited 3/26/13).

3. And, thankfully, in the limit when particles have very large energies, the relationship between their quantum wavelength and energy becomes very simple: $E = h/2\pi\lambda$ where λ is the wavelength and h is Planck's constant. This simple formula explains everything about the largest accelerators in the world, from the Fermilab Tevatron, to the SLAC Linac, to

LEP, to the Large Hadron Collider—in order to halve the size of λ we must double E—ergo high-energy particle accelerators are big.

4. Radio frequency cavities are another marvel of modern technology that was spun off from particle accelerator R&D. See "Microwave cavity," http://en.wikipedia.org/wiki/Microwave_cavity, "Klystron," http://en.wikipedia.org/wiki/Klystron (site last visited 3/26/13).

5. Widerøe's cavities were just spaces in a vacuum pipe between tubes of copper that were alternately charged plus and minus by an oscillating electric circuit. When the particles were between the tubes, they were in phase with an electric field (e.g., if an electron, the tube ahead would be charged positive, while the one behind would be negative). As they entered the tubes, the tubes changed polarity while the electrons merely drifted. See "Linear particle accelerator," http://en.wikipedia.org/wiki/Linear_particle_accelerator.

6. See the SLAC site: http://www6.slac.stanford.edu/research/ (site last visited 3/26/13).

7. "Fermilab Linac," http://www-ad.fnal.gov/proton/linac.html, and for the history: http://history.fnal.gov/linac.html (sites last visited 1/27/2013).

8. See, e.g., http://www.hep.ucl.ac.uk/~jpc/all/ulthesis/node15.html. See "International Linear Collider," http://en.wikipedia.org/wiki/International_Linear_Collider; see also http://www.linearcollider.org/ILC/GDE/Director%27s-Corner/2008/1-May-2008---The-art-of-decision-making---STF-phase-2-cavity-choice (sites last visited 3/26/2013).

9. See "Technetium-99," http://en.wikipedia.org/wiki/Technetium-99 (site last visited 3/26/2013).

10. "Magnetic Field Basics," http://www.physics4kids.com/files/elec_magneticfield.html (site last visited 3/26/2013). Digested from the article:

> Magnets and magnetism were known to ancients. The magnetic field on the surface of a spherical magnet was mapped using iron needles in 1269 by Petrus Peregrinus de Maricourt. He coined the term "poles" in analogy to Earth's poles where the field lines converged at two points on the sphere. In 1600 in a publication, *De Magnete*, William Gilbert of Colchester demonstrated explicitly that Earth is a magnet, helped to establish magnetism as a science. In 1819, Hans Christian Oersted discovered that an electric current generates a magnetic field encircling it. André-Marie Ampère in 1820 showed that parallel wires having currents in the same direction attract one another. Jean-Baptiste Biot and Félix Savart discovered the force law in 1820 which correctly predicts the magnetic field around any current-carrying wire.
>
> Extending these experiments, Ampère published his own successful model of magnetism in 1825. . . . Further, Ampère derived both Ampère's force law describing the force between two currents and Ampère's law which, like the Biot–

Savart law, correctly described the magnetic field generated by a steady current. Also in this work, Ampère introduced the term "electrodynamics" to describe the relationship between electricity and magnetism.

In 1831, Michael Faraday discovered electromagnetic induction when he found that a changing magnetic field generates an encircling electric field. He described this phenomenon in what is known as Faraday's law of induction. Later, Franz Ernst Neumann proved that, for a moving conductor in a magnetic field, induction is a consequence of Ampère's force law. In the process he introduced the magnetic vector potential which was later shown to be equivalent to the underlying mechanism proposed by Faraday. . . .

Between 1861 and 1865, James Clerk Maxwell developed and published Maxwell's equations which explained and united all of classical electricity and magnetism. The first set of these equations was published in a paper entitled *On Physical Lines of Force* in 1861. These equations were valid although incomplete. He completed Maxwell's set of equations in his later 1865 paper, *A Dynamical Theory of the Electromagnetic Field*, and demonstrated the fact that light is an electromagnetic wave. Heinrich Hertz experimentally confirmed this fact in 1887.

Although implicit in Ampère's force law the force due to a magnetic field on a moving electric charge was not correctly and explicitly stated until 1892 by Hendrik Lorentz who theoretically derived it from Maxwell's equations. With this last piece of the puzzle, the classical theory of electrodynamics was essentially complete.

Mapping the magnetic field of an object is simple in principle. First, measure the strength and direction of the magnetic field at a large number of locations. Then, mark each location with an arrow (called a vector) pointing in the direction of the local magnetic field with a length proportional to the strength of the magnetic field.

A simpler method to map the magnetic field is to "connect" the arrows to form magnetic field lines. On a magnetic field line diagram, the direction of the magnetic field at any point is represented by the direction of nearby field lines. Further, if drawn carefully, a higher density of nearby field lines indicates a stronger magnetic field.

Magnetic field lines are like the contour lines (constant altitude) on a topographic map in that a different mapping scale would show more or fewer lines. An advantage of using magnetic field lines, though, is that many laws of magnetism (and electromagnetism) can be stated completely and concisely using simple concepts such as the "number" of field lines through a surface. These concepts can be quickly "translated" to their mathematical form. For example, the number of field lines through a given surface is the surface integral of the magnetic field.

Various phenomena have the effect of "displaying" magnetic field lines as though the field lines are physical phenomena. For example, iron filings placed in a magnetic field line up to form lines that correspond to "field lines." Magnetic fields' "lines" are also visually displayed in polar auroras, in which plasma particle dipole interactions create visible streaks of light that line up with the local direction of Earth's magnetic field.

Field lines can be used as a qualitative tool to visualize magnetic forces. In ferromagnetic substances like iron and in plasmas, magnetic forces can be understood by imagining that the field lines exert a tension (like a rubber band) along their length, and a pressure perpendicular to their length on neighboring field lines. "Unlike" poles of magnets attract because they are linked by many field lines; "like" poles repel because their field lines do not meet, but run parallel, pushing on each other.

The twentieth century extended electrodynamics to include relativity and quantum mechanics. Albert Einstein, in his paper of 1905 that established relativity, showed that both the electric and magnetic fields are part of the same phenomena viewed from different reference frames.

11. Note that acceleration is the time rate of change of a velocity. Velocity is a vector, meaning it has both magnitude (speed) and direction. If the speed is constant but the direction varies in time, as is the case for uniform circular motion, then the particle is being accelerated.

12. See "André-Marie Ampère," http://en.wikipedia.org/wiki/Andr%C3%A9-Marie_Amp%C3%A8re (site last visited 3/26/2013).

13. This is called "direct current," or DC, not the "alternating current," or AC, which oscillates in the wires in your home; the oscillating current also produces magnetic fields, but the forces will oscillate and average to zero, so you won't see the wires deflecting one another

14. The magnetic property of iron is a consequence of the intrinsic spin of electrons in the iron atoms and is a rather complex quantum phenomenon. See "Ferromagnetism," http://en.wikipedia.org/wiki/Ferromagnetism (site last visited 1/26/2013).

15. Magnetic monopoles would in principle exist if certain conditions in particle physics were realized. Such objects would be extremely heavy (at least many TeV in mass, probably higher); there is to date no evidence, other than various theories, of their existence See "Magnetic monopole," http://en.wikipedia.org/wiki/Magnetic_monopole (site last visited 1/26/2013).

16. See iron filings revealing the magnetic field lines at http://en.wikipedia.org/wiki/Magnetism#History; see also http://www.wired.com/wiredscience/2011/09/magnetic-invisibility-cloak/ (sites last visited 3/26/2013), or search online for keywords "iron filings magnetic."

17. See "Cyclotron," http://en.wikipedia.org/wiki/Cyclotron and illustrations therein, (site last visited 1/26/2013).

18. See "Strong focusing," http://en.wikipedia.org/wiki/Strong_focusing and "Synchrotron," http://en.wikipedia.org/wiki/Synchrotron (sites last visited 1/26/2013).

19. "Synchrotron radiation," http://hyperphysics.phy-astr.gsu.edu/hbase/particles/synchrotron.html; http://www.hep.ucl.ac.uk/~jpc/all/ulthesis/node15.html; http://physik.uibk.ac.at/~emo/physics/synchrotron.html (sites last visited 1/26/2013).

20. See "Quadrupole magnet," http://en.wikipedia.org/wiki/Quadrupole_magnet (site last visited 1/26/2013).

21. The "FODO" lattice is discussed in D. Edwards and M. Syphers, *An Introduction to the Physics of High Energy Accelerators (Wiley Series in Beam Physics and Accelerator Technology)* (Wiley, 1992). For information on the Brookhaven AGS, see http://en.wikipedia.org/wiki/Alternating_Gradient_Synchrotron (site last visited 1/26/2013).

22. Lillian Hoddeson, Adrienne Kolb, and Catherine Westfall, *Fermilab: Physics, the Frontier, and Megascience* (University of Chicago Press, 2011); see also "Tevatron," http://en.wikipedia.org/wiki/Tevatron (sites last visited 3/26/2013). From the source:

> December 1, 1968 saw the breaking of ground for the linear accelerator (linac). The construction of the Main Accelerator Enclosure began on October 3, 1969 when the first shovel of earth was turned by Robert R. Wilson, NAL's director. This would become the 6.4 km circumference of Fermilab's Main Ring.
>
> The linac's first 200 MeV beam started on December 1, 1970. The booster's first 8 GeV beam was produced on May 20, 1971. On June 30, 1971, a proton beam was guided for the first time through the entire National Accelerator Laboratory accelerator system including the Main Ring. The beam was accelerated to only 7 GeV . . .
>
> A series of milestones saw acceleration rise to 20 GeV on January 22, 1972 to 53 GeV on February 4 and to 100 GeV on February 11. On March 1, 1972, the then NAL accelerator system accelerated for the first time a beam of protons to its design energy of 200 GeV. By the end of 1973, NAL's accelerator system operated routinely at 300 GeV.
>
> On 14 May, 1976 Fermilab took its protons all the way to 500 GeV. This achievement provided the opportunity to introduce a new energy scale, the Tera electron volt (TeV), equal to 1000 GeV. On 17 June of that year, the European Super Proton Synchrotron accelerator (SPS) had achieved an initial circulating proton beam (with no accelerating radio-frequency power) of only 400 GeV.
>
> The old copper magnet accelerator was shut down on August 15, 1977 for superconducting magnets to be mounted "piggy-back" on the main ring magnets. The "Energy Doubler," as it was known then, produced its first accelerated

beam—512 GeV—on July 3, 1983. Its initial energy of 800 GeV was achieved on February 16, 1984. On October 21, 1986 acceleration at the Tevatron was pushed to 900 GeV, providing a first proton–antiproton collision at 1.8 TeV on November 30, 1986.

The Main Injector, which replaced the Main Ring, was the most substantial addition, built over six years from 1993 at a cost of $290 million. Tevatron collider Run II began on March 1, 2001 after successful completion of that facility upgrade. From then, the beam had been capable of delivering an energy of 980 GeV.

On July 16, 2004 the Tevatron achieved a new peak luminosity, breaking the record previously held by the old European Intersecting Storage Rings (ISR) at CERN. That very Fermilab record was doubled on September 9, 2006, then a bit more than tripled on March 17, 2008, and ultimately multiplied by a factor of 4 over the previous 2004 record on April 16, 2010 (up to 4×10^{32} cm^{-2} s^{-1}).

By the end of 2011, the Large Hadron Collider (LHC) at CERN had achieved a luminosity almost ten times higher than Tevatron's (at 3.65×10^{33} cm^{-2} s^{-1}) and a beam energy of 3.5 TeV each (doing so since March 18, 2010), already ~3.6 times the capabilities of the Tevatron (at 0.98 TeV).

The initial design luminosity of the Tevatron was 10^{30} cm^{-2} s^{-1}, however the accelerator has following upgrades been able to deliver luminosities up to 4×10^{32} cm^{-2} s^{-1}.

The Booster is a small circular synchrotron, around which the protons pass up to 20,000 times to attain an energy of around 8 GeV. From the Booster the particles pass into the Main Injector, which was completed in 1999 to perform a number of tasks. It can accelerate protons up to 150 GeV; it can produce 120 GeV protons for antiproton creation; it can increase antiproton energy to 120 GeV, and it can inject protons or antiprotons into the Tevatron. The antiprotons are created by the Antiproton Source. 120 GeV protons are collided with a nickel target producing a range of particles including antiprotons which can be collected and stored in the accumulator ring. The ring can then pass the antiprotons to the Main Injector.

The Tevatron could accelerate the particles from the Main Injector up to 980 GeV. The protons and antiprotons are accelerated in opposite directions, crossing paths in the CDF and DØ detectors to collide at 1.96 TeV. To hold the particles on track the Tevatron uses 774 niobium-titanium superconducting dipole magnets cooled in liquid helium producing 4.2 teslas. The field ramps over about 20 seconds as the particles are accelerated. Another 240 NbTi quadrupole magnets are used to focus the beam.

The Tevatron confirmed the existence of several subatomic particles that were predicted by theoretical particle physics, or gave suggestions to their existence. In 1995, the CDF experiment and DØ experiment collaborations announced the discovery of the top quark, and by 2007 they measured its mass to a precision of nearly 1%. In 2006, the CDF collaboration reported the first measurement of B_s oscillations, and observation of two types of sigma baryons.

23. See "Tevatron," http://en.wikipedia.org/wiki/Tevatron (site last visited 6/21/13).

24. The antiprotons are created in a system called the "Antiproton Source." The Main Injector–accelerated 120 GeV protons collide with a nickel target, producing a spray of different subatomic particles, which includes antiprotons. These are collected and "cooled," meaning that compact stable bunches are formed. They are then stored in the accumulator ring, and then passed back to the Main Injector, then ultimately into the Tevatron.

25. See "Large Electron–Positron Collider," http://en.wikipedia.org/wiki/Large_Electron%E2%80%93Positron_Collider (site last visited 3/26/2013).

26. See "Large Hadron Collider," http://en.wikipedia.org/wiki/Large_Hadron_Collider (site last visited 3/26/2013).

27. See "Georges Charpak," http://en.wikipedia.org/wiki/Charpak (site last visited 3/26/2013).

28. See "International Linear Collider," http://en.wikipedia.org/wiki/International_Linear_Collider and http://www.linearcollider.org/ (sites last visited 3/26/2013).

CHAPTER 9. RARE PROCESSES

1. "Wilhelm Conrad Röntgen," http://www.nobelprize.org/nobel_prizes/physics/laureates/1901/rontgen-bio.html; see "Wilhelm Röntgen," http://en.wikipedia.org/wiki/Wilhelm_Röntgen; "The Discovery of X-Rays," http://www.ndt-ed.org/EducationResources/HighSchool/Radiography/discoveryxrays.htm (sites last visited 3/23/2013). Gottfried Landwehr, *Röntgen Centennial: X-Rays in Natural and Life Sciences*, ed. A. Hasse (Singapore: World Scientific Publishing Co., 1997), pp. 7–8.

2. See "Henri Becquerel," http://en.wikipedia.org/wiki/Henri_Becquerel. Quoting from the article:

> As often happens in science, radioactivity came close to being discovered nearly four decades earlier when, in 1857, Abel Niepce de Saint-Victor, who was investigating photography under Michel Eugène Chevreul, observed that uranium salts emitted radiation able to darken photographic emulsions. By 1861, Niepce de Saint-Victor realized that uranium salts produce a radiation that is invisible to our eyes. (Note that Niepce de Saint-Victor knew Edmond Becquerel, Henri Becquerel's father).

See "Beta decay," http://en.wikipedia.org/wiki/Beta_decay#History, "Marie Curie," http://en.wikipedia.org/wiki/Marie_Curie, and "Pierre Curie," http://en.wikipedia.org/wiki/Pierre_Curie (sites last visited 3/23/2013).

3. "What Is Radioactivity?" http://www.oasisllc.com/abgx/radioactivity.htm. For a good historical summary on radioactivity, see http://www.chemteam.info/Radioactivity/Disc-of-Alpha&Beta.html and http://www.chemteam.info/Radioactivity/Disc-Alpha&Beta-Particles.html (sites last visited 3/23/2013).

4. See "Ernest Rutherford," http://en.wikipedia.org/wiki/Ernest_Rutherford (site last visited 3/23/2013).

Half-life: In a certain time interval, called the "half life," there will be exactly half as much of a radioactive material as one began with; in another half-life interval the amount will be reduced to half as much again, or a quarter of the initial amount, and so on; materials with short half-lives are very radioactive, e.g., technecium-99 used in most medical imaging has a 2-hour half-life and therefore requires a small dose; ^{235}U, the weapons-grade isotope of uranium has a half-life of 700,000 years; ^{238}U about 4 billion years.

5. See "Positron emission tomography," http://en.wikipedia.org/wiki/Positron_emission_tomography and http://www.webmd.com/a-to-z-guides/positron-emission-tomography (sites last visited 3/23/2013).

6. See "Paul Dirac," http://en.wikipedia.org/wiki/Paul_Dirac (site last visited 3/23/2013). Dirac was one of quantum theory's towering figures. For one thing, Dirac wrote *the* book on quantum physics, called *The Principle of Quantum Mechanics*, which became the standard reference. Dirac's original contributions to quantum physics are among the greatest of the twentieth century, anticipating modern topology with his "magnetic monopole" and Feynman's "path integral" formulation of quantum theory. Perhaps one of the most profound discoveries of foundational physics in the twentieth century, however, happened when Dirac combined the theory of the electron with Einstein's theory of special relativity and discovered antimatter some four years before it was discovered in experiments. See also Graham Farmelo, *The Strangest Man: The Hidden Life of Paul Dirac, Mystic of the Atom* (Basic Books; First Trade Paper Edition, 2011).

7. The Pauli exclusion principle is discussed further in our book: *Quantum Physics for Poets* (Amherst, NY: Prometheus Books, 2011). The quantum orbital state of motion in an atom can actually have two electrons, but one must have spin up and the other spin down.

8. We call this view of the vacuum the "Dirac sea." The Dirac sea is not empty but rather is a completely filled "ocean" that metaphorically represents the infinity of filled negative energy levels.

Now, usually when a high-energy gamma ray collides with a negative-energy electron in the vacuum, nothing happens. A single gamma ray hitting a negative-energy electron cannot raise it out of the vacuum because such a process wouldn't conserve all of the necessary quantities that physics demands be conserved, i.e., momentum, energy, angular momentum. However, if there are other particles also participating in the collision (like a nearby heavy atomic nucleus to recoil slightly and conserve the overall momentum, energy, and angular momentum of the participants in the collision; we call this a 3-body collision), then the electron could be ejected out of the Dirac sea into a state of positive energy. The gamma ray could then successfully eject an electron out of its negative energy state and into one of positive energy, and that could register in the physicist's instruments.

9. See "Carl David Anderson," http://en.wikipedia.org/wiki/Carl_David_Anderson and "Positron," http://en.wikipedia.org/wiki/Positron (sites last visited 3/23/2013).

10. The momenta of the outgoing electron and proton didn't add up to that of the neutron either, since neutrons at rest in the lab decayed into protons and electrons that were not seen to be emitted back-to-back; the reaction can also be "crossed" into $p^+ + e^- \rightarrow n+$ (missing energy).

11. See "Wolfgang Pauli," http://en.wikipedia.org/wiki/Wolfgang_Pauli. The full letter is in the CERN Pauli Archive, which may be visited online at www.library.cern.ch/archives/pauli/paulileter.html and is reproduced in our book *Symmetry and the Beautiful Universe* (Amherst, NY: Prometheus Books, 2007). We thank the CERN Pauli Archive Committee for permission to reproduce it.

12. It also allows the angular momentum to be conserved if the neutrino has spin 1/2; Enrico Fermi gave the name "neutrino" to the invisible particle in the decay process. Pauli had used the name "neutron" for his new particle, the name we now use for the heavy neutral constituent of the nucleus. You will note that this is a slight variation of the process $p^+ + e^- \rightarrow n^0 + \nu_e$, which causes a supernova. The squeezing together of a proton and electron can happen only at the extreme densities inside a massive star collapse. A neutron in free space decays into proton, electron, and neutrino a half-life of about 11 minutes by the related process of "beta decay," $n^0 \rightarrow p^+ + e^- + \bar{\nu}_e$.

13. Fermi's theory involves an "interaction strength" that is G_F. If we use the units in which $\hbar = c = 1$, then we find 175 GeV $= 1/\sqrt{2\sqrt{2}G_F}$.

14. See "Frederick Reines," http://en.wikipedia.org/wiki/Frederick_Reines and "Clyde Cowan," http://en.wikipedia.org/wiki/Clyde_Cowan (sites last visited 3/26/2013).

15. For the Lederman, Schwartz, Steinberger experiment, see chap. 6, note 1. When a third heavy lepton was discovered, the τ, or "tau," it was quickly realized that there was also a τ-neutrino. Today we know that there are three distinct kinds of neutrinos, electron neutrinos (ν_e), muon neutrinos (ν_μ), and tau neutrinos (ν_τ), each matching their charged lepton counterparts. This zoology of particles is further elaborated in the Appendix. The τ-neutrino was discovered at Fermilab in 2000, in the "DONUT experiment"; see "Tau neutrino," http://en.wikipedia.org/wiki/Tau_neutrino.

16. "Clue," or "Cluedo," is great preparation for a budding physicist. See http://en.wikipedia.org/wiki/Cluedo (site last visited 3/26/2013).

17. Lederman and Hill, "*Symmetry and the Beautiful Universe*," chap 8, note 2; T. D. Lee and C. N. Yang, "Question of Parity Conservation in Weak Interactions," *Physical Review* 104 (1956); J. Bernstein, "Profiles: A Question of Parity," *New Yorker Magazine* 38 (1962); M. Gardner, *The New Ambidextrous Universe: Symmetry and Asymmetry, from Mirror Reflections to Superstrings* (New York: W. H. Freeman and Co., 1991).

18. Even a complex break shot, where a cue ball scatters the ten pool balls, could in principle occur in a time-reversed way, but it is simply extremely improbable that it could be arranged to occur—this is why complex systems look funny when they run backward in time—the reversed physical processes are allowed, but it would be virtually impossible to set everything in motion to make things happen that way. The evolution of life is such a process—it is governed by the laws of physics, but it happens very slowly over many, many "collisions" of large macromolecules under rare circumstances. Once certain self-replicating macromolecules are formed, the process can build greater complexity through "random selection."

19. In physics we always pose, and solve, "if-then" problems. Let's consider the following elementary physics question (Q1): *If* a particle at time t_1 is located at x_1 traveling at a velocity, V, *then* where will the particle be at time t_2? The answer is $x_2 = x_1 + V(t_2 - t_1)$. But even this simple result illustrates some deep philosophical issues as to how we describe nature.

Consider now *a time-reversed question* (Q2): "*If* at time t_1 the particle is located at x_2 and traveling with velocity $-V$ (velocities change sign when we reverse the direction of time, as you well know by running a DVD backward and seeing a car driving in reverse down the highway), *then* where will it be at time t_2?" Now the answer, by common sense, must be x_1. And indeed, we see, upon a little rearranging, that our previous formula gives us $x_1 = x_2 - V(t_2 - t_1)$.

This is indeed the correct answer for the time-reversed question, yet it came from the original problem's solution after a little rearranging of the math. The answer for the forward-in-time question contains the answer for the backward-in-time question—we get both from one in the same physics equation! Our physical description of this system is the same if time is running forward as when time is running backward. In the second question, Q2, we set up *initial conditions* that were the *opposite* to those in the first question, Q1, that is, in Q2 we put the particle at the location, x_2, where it ended up in Q1, and we reversed the direction of motion, replacing V by $-V$. We find that after an equivalent time interval, the particle in Q2 gets to location x_1, where it started in Q1. This shows that we can do time-reversed physics without actually reversing the flow of time. We need only reverse the directions of motions and swap the final destinations for the initial one.

20. The particle K^0 and the antiparticle $\overline{K}^0$ actually oscillate back and forth between one another, $K^0 \leftrightarrow \overline{K}^0$. If CP is an exact symmetry, then the oscillation phase from the K^0 into the $\overline{K}^0$ should be *exactly the same* as the reverse oscillation phase, from the $\overline{K}^0$ into

the K^0. Experimentally, however, it is found that the oscillation phase to go from $K^0 \leftrightarrow \bar{K}^0$ is slightly different, at one part in a thousand, than the oscillation phase from $\bar{K}^0 \leftrightarrow K^0$. J. H. Christenson, J. W. Cronin, V. L. Fitch, and R. Turlay, "Evidence for the 2π Decay of the K(2)0 Meson," *Physical Review Letters* 13 (1964): 138–40. This is not CP invariant. In refined experiments with neutral K-mesons, direct confirmation of the violation of T-symmetry has also been confirmed. The combined symmetry transformation, CPT, is a symmetry of the decays. CP violation is now turning up in other particles, called B-mesons, containing the heavy beauty quark. Search online for "CP violation."

CHAPTER 10. NEUTRINOS

1. See "Supernova" and references therein, http://en.wikipedia.org/wiki/Supernova (site last visited 3/28/13). Or search online for "supernova" for some great images. We are mainly describing Type II supernovae in the text. There are several kinds of supernovae with differing evolutionary processes, as detailed in the article.

2. See "Galaxy," http://en.wikipedia.org/wiki/Galaxy (site last visited 3/28/13). Search online for "galaxies" for images.

3. See "Eta Carinae," http://en.wikipedia.org/wiki/Eta_carinae (site last visited 3/28/13). Under the heading "Possible Effects on Earth," "It is possible that the Eta Carinae hypernova or supernova, when it occurs, could affect Earth, which is about 7,500 light years from the star. . . . a certain few [scientists] claim that radiation damage to the upper atmosphere would have catastrophic effects as well. At least one scientist has claimed that when the star explodes, 'it would be so bright that you would see it during the day, and you could even read a book by its light at night.'" A supernova or hypernova produced by Eta Carinae would probably eject a gamma ray burst (GRB) out from both polar areas of its rotational axis. Calculations show that the deposited energy of such a GRB striking Earth's atmosphere would be equivalent to one kiloton of TNT per square kilometer over the entire hemisphere facing the star, with ionizing radiation depositing ten times the lethal whole body dose to the surface. This catastrophic burst would probably not hit Earth though, because the rotation axis does not currently point toward our solar system. If Eta Carinae is a binary system, this may affect the future intensity and orientation of the supernova explosion that it produces, depending on the circumstances.

Our advice is that you hug your children and call your Mom and tell her you love her.

4. See "Neutrino oscillation," http://en.wikipedia.org/wiki/Neutrino_oscillation (site last visited 3/28/13).

5. Recall from chapter 6: The L electron carries a weak charge, and R does not. Weak charge must also be conserved, but the Higgs field has filled the vacuum with a large reservoir of weak charge. So, L can convert to R by dumping its weak charge into the Higgs field

reservoir; and R can convert to L by absorbing the weak charge back again out of the Higgs field reservoir, making L. But the Higgs field does not absorb or relinquish *electric charge*, so the L electron must have the same charge as the R electron; the R positron must have the same charge as the L positron.

6. See "Ettore Majorana," http://en.wikipedia.org/wiki/Ettore_Majorana (site last visited 3/26/2013). The term "Majorana particle" is now commonly used but is erroneous, because the particle is actually one with a "Majorana mass." The term "Majorana particle" was historically reserved for spin-1/2 particles whose wave functions are real, which can only occur in special space-time dimensionalities, like 2, 6, 8, 10, etc.

7. See "Neutrinoless double-beta decay," http://en.wikipedia.org/wiki/Double_beta_decay#Neutrinoless_double-beta_decay (site last visited 3/26/2013).

8. This is called the "neutrino seesaw mechanism," http://en.wikipedia.org/wiki/Seesaw_mechanism (site last visited 3/26/2013).

9. See "Leptogenesis (physics)," http://en.wikipedia.org/wiki/Leptogenesis_%28physics%29 (site last visited 3/26/2013). Quoting from the source:

> The Big Bang produced matter and antimatter directly, but in nearly equal amounts. Today, however, we see no antimatter left in the universe. The cosmic annihilation of matter and antimatter should have been almost complete, leaving not nearly enough leftover matter to form the billions of stars that we see today, and us. Where did all this matter come from? Or, where did all the antimatter go? The process of leptogenesis could be the answer. Neutrinos are very different from other kinds of matter, and may be the only matter particles that are their own antiparticles. Neutrinos also have very tiny masses which suggests that the origin of neutrino mass involves much shorter-range interactions with hypothesized superheavy neutrinos. This may provide an experimental window to leptogenesis. When theorists rerun the tape of the Big Bang introducing superheavy partner neutrinos with nonstandard CP symmetry, the result is leptogenesis. The heavy neutrinos fall apart into light neutrinos, producing an excess of matter over antimatter. In the hot environment of the early universe, this excess is quickly passed along to all the particles that we are made of. If the theory of leptogenesis is correct, we owe our existence to neutrinos from the big bang.

10. See "Bruno Pontecorvo," http://en.wikipedia.org/wiki/Bruno_Pontecorvo (site last visited 3/26/2013).

11. See "Raymond Davis," http://en.wikipedia.org/wiki/Raymond_Davis_Jr. and "Homestake experiment," http://en.wikipedia.org/wiki/Davis_Experiment (sites last visited 3/26/2013).

12. See "Masatoshi Koshiba," http://en.wikipedia.org/wiki/Masatoshi_Koshiba (site last visited 3/26/2013).

13. See "Super-Kamiokande," http://en.wikipedia.org/wiki/Super_Kamiokande (site last visited 3/26/2013).

14. "How Does NOνA Work?" http://www-nova.fnal.gov/how-nova-works.html. See "Neutrino oscillation," http://en.wikipedia.org/wiki/Neutrino_oscillation (sites last visited 3/26/2013). It's important to realize that we did not dig a 500-mile-long tunnel from Fermilab to Soudan, Minnesota—the neutrinos propagate freely and unimpeded through the earth under Wisconsin (Go, Packers!).

15. Fermilab is playing an active role in this development with a trial liquid argon detector experiment called MicroBooNE: "ArgoNeuT," http://www.fnal.gov/pub/science/experiments/intensity/argoneut.html; "MicroBooNE," http://www-microboone.fnal.gov/ (sites last visited 3/26/2013). Liquid argon-based time-projection chambers are also under active study for dark matter and neutrinoless double-beta decay detection, addressing the issue of Majorana vs. Dirac neutrino masses.

16. See "Proton decay," http://en.wikipedia.org/wiki/Proton_decay (site last visited 3/26/2013).

CHAPTER 11. PROJECT X

1. "The Shutdown Process," http://www.fnal.gov/pub/tevatron/shutdown-process.html (site last visited 3/26/2013).

2. John Matson, "Life after Tevatron: Fermilab Still Kicking Even Though It Is No Longer Top Gun," *Scientific American* (January 2012), http://blogs.scientificamerican.com/observations/2012/01/31/life-after-tevatron-fermilab-still-kicking-even-though-it-is-no-longer-top-gun/ (site last visited 1/23/2013). The lessons from the Tevatron are interesting. The top quark was initially thought to be light, about 90 GeV, and it should then have been seen in the first year of running. However, the top quark has turned out to have a very large mass of 172 GeV, and this required several years of patient searching until evidence for it finally emerged at the Tevatron, followed by a bona fide discovery in 1995. Had the top quark been about 60 GeV heavier we may never have found it at the Tevatron.

It was estimated in 1991 that, if the Higgs was lighter than about 140 GeV, the Tevatron could see it with 30 inverse femtobarns of data. The Tevatron luminosity increased significantly, due to the heroic efforts of the Fermilab Accelerator Division, to the point that it became clear the Higgs could have been seen with a concerted effort by the lab. The prediction for the required luminosity turned out to be right on the nose, but unfortunately the Tevatron ended operations, having delivered one-third of the required Higgs discovery's integrated luminosity. Still, the decay mode, by which there is now some evidence of the Higgs boson at the Tevatron, is the decay of Higgs into a b quark + anti–b quark final state. This decay mode is very important to our understanding of the Higgs properties. But this mode, and other "matter decay modes," of the Higgs will be established in the all-important LHC run, due to commence January 1, 2015.

3. See Fermilab's Project X website: http://projectx.fnal.gov/; in particular, one can access the "Project X Book" at this site, which gives comprehensive literature on the experimental program and machine and detector designs. See also "Project X (accelerator)," http://en.wikipedia.org/wiki/Project_X_%28accelerator%29; "Fermilab's Project X Could Offer Potential Energy Applications," http://www.symmetrymagazine.org/breaking/2011/04/12/fermilabs-project-x-may-have-a-potential-energy-application (sites last visited 1/23/2013).

4. "Muon Storage Ring": http://www.cap.bnl.gov/mumu/studyii/final_draft/chapter-7/chapter-7.pdf; "Muon Ring Could Act as a Neutrino Factory," http://cerncourier.com/cws/article/cern/28043 (sites last visited 1/23/2013).

5. "The E821 Muon (g-2) Home Page," http://www.g-2.bnl.gov/; "Muon g-2," http://muon-g-2.fnal.gov/ (sites last visited 6/24/2013).

6. See "Neutrino Factory," http://en.wikipedia.org/wiki/Neutrino_Factory (site last visited 4/3/2013).

7. See "Muon collider," http://en.wikipedia.org/wiki/Muon_Collider (site last visited 4/3/2013).

8. The charged-kaon decay mode has been previously studied by the Brookhaven E787/949 experiment using a high- intensity stopped-kaon technique to yield a total of seven candidate signal events. The NA62 experiment at CERN is currently pursuing a new in-flight technique with the aim of achieving a 100-event sensitivity at the Standard Model level. The process $K_L \rightarrow \pi \nu \nu$ is a purely CP-violating process, that is predicted in the Standard Model theoretically at the 1 percent level of precision. The observation and precise measurement of this rare process will constitute a major triumph in kaon physics with the potential of discovering discrepancies. The KOTO experiment at J-PARC in Japan is pursuing a staged approach to reach single-event sensitivity, with an ultimate goal of reaching 100-event sensitivity, at the Standard Model level. This establishes the need for a multi-1,000-event future Project-X-based experiment.

9. See "Quantum electrodynamics," http://en.wikipedia.org/wiki/Quantum_Electrodynamics, "Richard Feynman," http://en.wikipedia.org/wiki/Richard_Feynman, "Julian Schwinger," http://en.wikipedia.org/wiki/Julian_Schwinger, "Sin-Itiro Tomonaga," http://en.wikipedia.org/wiki/Sin-Itiro_Tomonaga (sites last visited 4/3/2013).

10. We say that electric fields are vectors (they reflect like velocities or position vectors in mirrors); magnetic fields are "pseudo-vectors" and reflect with an opposite sign in mirrors. One has to be a little careful with this, because with vectors it depends upon the orientation of the system and of the mirror.

11. Also, it should be noted that EDM experiments provide very sensitive limits on the existence of electric dipoles and have already bitten the theorists in the pants. The

current upper limit on the existence of the electron EDM is about 10^{-27} e-cm coming from studying "polar molecules" like Yb-F (see, e.g., "A New Upper Limit on the Electron's Electric Dipole Moment," *Physics Today* 12 [August 2011]). This result already severely constrained the Minimal Supersymmetric Standard Model (MSSM) model, before the LHC arrived on the scene. We believe that supersymmetry has a very good chance of being true, but perhaps only at extremely high and inaccessible energy scales, and perhaps in a novel form that no one has yet conceived of.

12. See "Electron electric dipole moment," http://en.wikipedia.org/wiki/Electron_electric_dipole_moment, "Neutron electric dipole moment," http://en.wikipedia.org/wiki/Neutron_electric_dipole_moment (sites last visited 4/3/2013).

13. See "Proton therapy," http://en.wikipedia.org/wiki/Proton_therapy (sites last visited 4/7/2013).

14. "Accelerator Driven Subcritical Reactors," http://www.academia.edu/1684005/Accelerator_Driven_Subcritical_Reactors; "Accelerator-Driven Nuclear Energy," http://www.world-nuclear.org/info/Current-and-Future-Generation/Accelerator-driven-Nuclear-Energy/#.UVX7efKbFXs (sites last visited 4/3/2013). Much more can be found by searching online for "accelerator driven subcritical reactors." See also R. P. Johnson et al., "GEM*STAR—New Nuclear Technology to Produce Inexpensive Diesel Fuel from Natural Gas and Carbon," *Proceedings of IPAC2013*, Shanghai, China.

15. "Muon Accelerator Program," http://map.fnal.gov/ (site last visited 4/7/2013).

CHAPTER 12. BEYOND THE HIGGS BOSON

1. For in introduction to the cosmological theory, see Steven Weinberg, *The First Three Minutes* (New York: Basic Books, 1977). Search online for "cosmology" and "big bang" for various wiki articles.

2. See "*Nimitz*-class aircraft carrier," http://en.wikipedia.org/wiki/Nimitz-class_aircraft_carrier, "*Gerald R. Ford*–class aircraft carrier," http://en.wikipedia.org/wiki/Gerald_R._Ford-class_aircraft_carrier; Ronald O'Rourke, "Navy CVN-21 Aircraft Carrier Program: Background and Issues for Congress," http://www.history.navy.mil/library/online/navycvn21.htm (sites last visited 4/7/2013).

3. *Estimated* shale oil alone is 1.5 trillion barrels, which at $100/bbl is $150 trillion. See, e.g., http://dailyreckoning.com/oil-shale-reserves/; see also http://abcnews.go.com/Business/american-oil-find-holds-oil-opec/story?id=17536852#.UVcVQPKbFXs. Note the word "estimated" is key here, since proven reserves are significantly less and don't include shale. Estimated coal and natural gas reserves are comparable (sites last visited 4/7/2013).

4. See, e.g., http://rutledgecapital.com/2009/05/24/total-assets-of-the-us-economy-188-trillion-134xgdp/. This article essentially asks the sensible question "Why are we

sweating the $17 trillion debt when we have $200 trillion in assets?" (site last visited 4/7/2013).

5. See http://www.forbes.com/forbes-400/list/ (site last visited 4/7/2013).

6. "Conjecture on the Physical Implications of the Scale Anomaly," invited talk on the occasion of the 75th birthday celebration of Murray Gell-Mann ("Murraypalooza"), Christopher T. Hill (2005), downloadable pdf file at http://arxiv.org/pdf/hep-th/0510177.pdf and references therein.

7. Steven Weinberg, "The Crisis of Big Science," *New York Times Book Review*, May 10, 2012, http://www.nybooks.com/articles/archives/2012/may/10/crisis-big-science/ (site last visited 3/8/2013).

APPENDIX

1. See "Quark," http://en.wikipedia.org/wiki/Quark, "Murray Gell-Mann," http://en.wikipedia.org/wiki/Gell-Mann, "George Zweig," http://en.wikipedia.org/wiki/George_Zweig (sites last visited 3/13/13). When Gell-Mann proposed the term "quark," borrowed from the passage in James Joyce's novel *Finnegans Wake*: "Three quarks for Muster Mark," he thankfully broke the tradition that everything requires a Greek alphabetical symbol for nomenclature in particle physics. The idea of quarks was also independently proposed by George Zweig, a colleague of Gell-Mann at Caltech who happened to be on a visit to CERN and wrote down the idea in a famous unpublished CERN preprint. Zweig chose the name "aces." Zweig realized that certain dynamical properties of the many newly discovered particles could be explained on the basis of the next layer of matter, the quarks.

2. See "James D. Bjorken," http://en.wikipedia.org/wiki/Bjorken, "Deep inelastic scattering," http://en.wikipedia.org/wiki/Deep_inelastic_scattering (sites last visited 3/13/13).

3. See "Quantum chromodynamics," http://en.wikipedia.org/wiki/Quantum_chromodynamics; Frank Wilczek, "QCD Made Simple," http://www.frankwilczek.com/Wilczek_Easy_Pieces/298_QCD_Made_Simple.pdf (sites last visited 3/13/13).

4. See "Generation (particle physics)," http://en.wikipedia.org/wiki/Generation_%28particle_physics%29 (site last visited 3/13/13). We actually have some understanding of the fact that the charges of a generation must add to zero because of the cancelation of the Adler-Bardeen-Bell-Jackiw anomalies, a quantum threat to the symmetries in the weak interactions that must be perfectly canceled to zero. We also have beautiful "unified theories," such as the Georgi–Glashow SU(5) theory that "predicts" this pattern. However, in any theory, we can't be absolutely sure that the electron goes with the u and d quarks, as opposed to the t and b quarks, or some other scrambling of things.

The CP violation observed in quarks *does require* all three generations, for technical reasons, and we also know that some kind of CP violation is necessary for matter to exist in the universe at all. Also, all quarks and leptons are active in the early universe and play a role in the formation of the universe we end up with.

5. The number of gluons is 8 = 9 – 1. 9 is the number of (color, anti-color) pairs that we can ever possibly have. One combination, $r\overline{r} + b\overline{b} + g\overline{g}$, is not an SU(3) group element. That is, it doesn't rotate anything in color space and doesn't arise as a gluon.

6. See also "Spin (physics)," http://en.wikipedia.org/wiki/Spin_%28physics%29 (site last visited 3/13/13).

7. See "Wave function," http://en.wikipedia.org/wiki/Wave_function (site last visited 3/13/13).

8. See "Boson," http://en.wikipedia.org/wiki/Bosons, "Fermion," http://en.wikipedia.org/wiki/Fermion (sites last visited 3/13/13).

9. See "James Clerk Maxwell," http://en.wikipedia.org/wiki/James_Clerk_Maxwell (site last visited 3/13/13). Maxwell, Scottish born and living only to age 48, is a towering figure in the history of science. His importance in the history of physics is comparable to that of Einstein and Newton. He was the first to recognize that light is a propagating wave disturbance of electric and magnetic fields, and was responsible for finding a solution to the equations that describe all electric and magnetic phenomena, known as Maxwell's equations. The laws of special relativity are already contained in Maxwell's theory—Einstein unearthed them by contemplating the symmetries of the equations under different states of inertial motion.

10. J. D. Jackson and L. B. Okun, "Historical Roots of Gauge Invariance," *Reviews of Modern Physics* 73 (2001): 663–80; John P. Ralston (private communication); Jackson and Okun write:

> Notable in this regard, but somewhat peripheral to our history of gauge invariance, was James MacCullagh's early development of a phenomenological theory of light as disturbances propagating in a novel form of the elastic ether, with the potential energy depending not on compression and distortion but only on local rotation of the medium in order to make the light vibrations purely transverse. . . . MacCullagh's equations correspond (when interpreted properly) to Maxwell's equations for free fields in anisotropic media. We thank John P. Ralston for making available his unpublished manuscript on MacCullagh's work.
>
> Thus, MacCullagh actually constructed a theory of light as a propagating wave disturbance in a material medium, an "ether," in 1839. This theory is equivalent to Maxwell's theory, of some 25 years later, and it involves the concept of an unobservable gauge field, hence MacCullagh seemed to have understood the symmetry principle that is required. But the connection of the underlying physical picture here, involving the concept of twists, or local rotations, in a material medium, to electrodynamics is remote. This discovery has gone almost completely

unnoticed. MacCullagh, whose relationships with the rest of the physics community were not happy ones, and whose life ended tragically in suicide, may have been a man too far ahead of his time.

11. See "Gauge theory," http://en.wikipedia.org/wiki/Gauge_theory, http://en.wikipedia.org/wiki/Introduction_to_gauge_theory, "Yang–Mills theory," http://en.wikipedia.org/wiki/Yang%E2%80%93Mills_theory (sites last visited 3/13/13). More mathematically, we can describe the gauge invariance of electrodynamics: We consider a *complex phase factor*, which is just an exponential, like $e^{i\theta}$, where θ is real, and this has a magnitude of unity, i.e., $1 = | e^{i\theta} |^2$. We also consider the electron wave function, which is a complex valued function of space and time, $\psi(\vec{x}, t)$. Multiplying the electron wave function by this factor means $\psi(\vec{x},t) \rightarrow e^{i\theta}\psi(\vec{x},t)$ doesn't change the magnitude of the electron's wave function, and it therefore shouldn't affect the measured probabilities $| \psi(\vec{x},t)|$.

12. See "Feynman diagrams," http://en.wikipedia.org/wiki/Feynman_diagrams (site last visited 3/13/13).

13. See "Yang–Mills Theory," http://en.wikipedia.org/wiki/Yang-Mills; see also "Special unitary group," under "n = 2," http://en.wikipedia.org/wiki/SU%282%29#n_.3D_2 (sites last visited 3/13/13).

14. See "Standard Model," http://en.wikipedia.org/wiki/Standard_Model (site last visited 3/13/13).

INDEX

PERSONALITY
a scientific approach

This book was originally published
by Appleton-Century-Crofts in its *Century Psychology Series*

Kenneth MacCorquodale
Gardner Lindzey
Kenneth E. Clark

EDITORS